MASTERING

ELECTRONICS

MASTERING

ELECTRONICS

Latest Technology

SECOND EDITION

JOHN WATSON

McGraw-Hill Book Company
New York St. Louis San Francisco Montreal Toronto

Library of Congress Cataloging-in-Publication Data

Watson, John (John Richard)
  Mastering electronics.

  Bibliography: p.
  Includes index.
  1. Electronics.   I. Title
TK7816.W35   1986   621.381   86-20122
ISBN 0-07-068480-4

  4567890   DOC/DOC   89321098

ISBN 0-07-068480-4

First published in Great Britain by
Macmillan Education Ltd. in 1983.

First American edition published by McGraw-Hill in 1986.

Printed and bound by R.R. Donnelley & Sons Company.

This one, at last, is for Olly

CONTENTS

CONTENTS

III. DIGITAL ELECTRONICS

CONTENTS

PREFACE

Mastering electronics with one book is a tall order.

I set out to write this preface as a rationale for the book, and as an explanation of the reasons why I wrote it this way. I was going to describe the electronics industry and the way it has grown almost beyond recognition in the last couple of decades. I was going to say how the changes in the technology have resulted in changes in the way the subject is—must be—taught. But, well into the third page, I decided not to. Much of it is in the first and the final chapters, anyway.

Instead, I will simply (and briefly) explain what I have done. *Mastering Electronics* is intended as an introduction to the subject for anyone who wants to understand the basics of most areas of electronics. I have tried to make the coverage as broad as possible within the confines of an affordable book. Most aspects of electronics—from basic semiconductor theory to television and computers—have been fitted in, but with deliberately unequal coverage. In what is intended to be a fairly basic book, it seems sensible to devote proportionately more space to the more fundamental topics.

On the principle that if a picture is worth a thousand words, a circuit diagram must be worth five thousand, I have been lavish with the illustrations. The concomitant is that the text is, in places, on the dense side and will repay rather careful reading.

I have written the book in a way that reflects modern thinking and requirements for the electronics technologist, placing an emphasis on systems and on electronics in 'real life' rather than in the laboratory. Following the structure of most recent syllabuses, I have reduced the number of mathematical descriptions to an absolute minimum, including only such formulae as are essential for calculations. This has saved a certain amount of space, which I have used to put in suggestions for practical circuits and experiments. Wherever I recommend a circuit for practical work, I have built and tested the design before committing it to paper.

Mastering Electronics can be used as a self-teaching book or as a textbook; I think that on balance it has probably gained something in being designed for this dual role.

It appears to be usual to use the end of the preface to thank everybody who helped me produce the book. I think I will apologise instead, to my wife and boys, for a certain degree of preoccupation during the last few months. . . .

JOHN WATSON

SYMBOLS AND UNITS

Successive attempts to 'metricate', both in the UK and in the USA, have left the electronics industry a little confused about units in some areas. Similarly, different 'standards' have been issued in different countries regarding the symbols to be used in circuit and logic diagrams, and although there are general similarities, there are disagreements about the details.

I have tried to take a middle and sensible course in *Mastering Electronics*. I have used SI metric units for all measurements, except where the original is clearly in Imperial units, imported, paradoxically, from the USA. For example, the plastic DIL pack (dual in-line) for integrated circuits has a standard spacing between connecting pins: it seems silly to assert that the spacing is 2.54 mm, when it is clearly $\frac{1}{10}$ inch!

I have used British Standard symbols in all circuit and logic diagrams except where, for reasons of its own, the electronics industry has obstinately refused to use them. In such cases I have bowed to the majority opinion and done what everybody else does. Where symbols are distinctly different, for example BS 3939:1985/IEC 617-12: 1983 and ANSI Y32.14: 1973 standard logic symbols, I have shown both and then stuck with the BS/IEC version.

For component values, I have generally omitted the units in circuit diagrams; thus a 1.8 kΩ resistor becomes 1.8 k on the diagram. I have avoided the 1k8 convention.

EUROPEAN TRANSISTOR TYPE NUMBERS

European transistor types have been used in the many of the circuits given in this book. In almost all cases, the exact specification is not important, and the experienced constructor will be able to make an educated guess at the device required. However, the following table of direct equivalents (including equivalents of obsolete types) may be useful.

BC109	2N930	BC143	2N4037
BC107	2N929	AC141	2N2430
BC479	2N3965	AC142	2N2706
BFX51	2N731	OC71	2N406
BFX88	2N2905A	OC28	2N1533
BC142	2N2297	AC128	2N2706

MASTERING

ELECTRONICS

INTRODUCTION TO
ELECTRONICS

Only a few years ago, you would not have found the study of 'electronics' on the curriculum of any university or college. The subject was not considered important enough to be given a category of its own, and was studied as a rather specialist branch of electricity. Today, this relatively new branch of technology has made an impact on almost every aspect of our lives. Electronics has made possible a range of inventions, from televisions to spacecraft. Many of these inventions—or the effects of them—have revolutionised the way we live.

The growth of electronics as a branch of technology has been unprecedentedly rapid. Never before has such a completely new technology been developed so quickly or so effectively, and so universally. Electronic techniques are now applied to all branches of science and engineering. A study of electronics is therefore central to any science or engineering course, and more and more people, in all walks of life, are going to be needing a basic knowledge of electronic technology.

It is true that electronics developed from the study of electricity. Early ideas about the way electric current could flow through conductors and through a vacuum led to the development of useful radio systems. It was possible to send messages around the world using what was, by today's standards, incredibly simple and crude equipment. The Second World War provided an urgent requirement for more sophisticated communication and other electronic systems. The invention of radar (radio direction finding and ranging) required a big step forward in theory and an even bigger step forward in engineering. The study of electronics gradually became an important subject in its own right, and the radio engineer became a specialised technician.

The development of television led to one of the most massive social changes that has taken place. Many households now possessed televisions, radios and record players. In some branches of industry, electronic systems became useful—but electronic devices not directly concerned with the wireless transmission of sound or pictures were still something of a curiosity.

Only in the early 1960s did electronics technology 'come of age' thanks to the work of three scientists, Bardeen, Brattain and Shockley, in the Bell laboratories in the USA. In 1957 they assembled the first working transistor.

1.1 THE MICROELECTRONIC REVOLUTION

In order to understand just how much impact the invention of the transistor was to have, it is necessary to realise that every electronic machine required the use of *valves*. Valves will be described in Chapter 5, but for the time being it is enough to know that valves are rather inconvenient devices for handling electrons. Valves are rather large, difficult to produce on a large scale, and extremely wasteful of power. The transistor, on the other hand, can be made quite small and uses relatively little power. It is much cheaper to make than a valve and lends itself to mass-production techniques. More importantly, the transistor can be made *very* small. There is a theoretical limit to the size of a transistor, but this limit is astonishingly tiny. Using transistor technology it became possible to make electronic circuits very complicated and very small. It is quite difficult to get an idea of the difference in scale between electronic devices using valves and electronic devices using microelectronic technology. It does not give a very valid comparison to look at a radio receiver of the 1930s and compare its size with a modern transistor radio. The size of the transistor radio is not limited by technology but rather by the size of the person who is expected to operate it. Transistor radios can be made extremely small, but the lower limit of smallness is reached when it becomes impossible to operate the controls!

A more useful comparison can be made by looking at computers. One of the first working, large-scale computers was made in the late 1940s. It occupied an area about equal to that of an hotel suite, and used as much power as a medium-sized street of houses. It was vastly expensive, and vastly unreliable—on average a valve had to be changed once every ten minutes.

I am fortunate enough to own a computer of substantially more power than that of the giant of the 1940s. It cost me less than one week's pay, and fits neatly into my jacket pocket. It will operate for approximately 300 hours on two tiny batteries and I would be surprised if it ever went wrong.

All this gives an indication of the scale of change that has taken place as a result of the development of electronics, and of the 'microelectronic revolution' sparked off by the invention of the transistor.

1.2 **ELECTRONICS TECHNOLOGY TODAY**

Most of the development in electronics has taken place since 1960. Inevitably there have been large changes in the way the subject is taught, and indeed in what is taught. Solid-state technology has even led to a change in the way we look at the physics of the atom. Models of the atom which work well for electronics based on the relatively crude 'valve' technology are often not sufficient for a useful description of modern microelectronic components.

This book is divided into three parts. Part I is a revision of **basic electricity**. This part of the book goes rather faster than the other two parts. It will provide you with sufficient knowledge of electricity to enable you to understand the rest of the book. It is not, however, a complete study of the subject but will be a useful memory-jogger for anyone who has studied electricity at school but who needs to be reminded of what was learnt there.

Part II deals with **linear electronics**. Broadly, this is a study of electronic components and systems that deal with continuously varying quantities and represent these quantities by means of an electrical analogue. Electronic systems that involve this technology include television and radio, audio and video reproduction equipment, and a number of electronic instruments.

Part III deals with **digital electronics**. In digital electronics, systems deal only with numbers, usually represented by the presence or absence of an electrical signal. Digital electronic systems include, of course, computers and calculators, and a number of electronic instruments. Digital electronics is probably the most rapidly developing branch of electronics technology. Various functions previously carried out by linear systems are now being dealt with digitally—one major example is studio-quality audio recording and reproduction. A sound signal can now be represented as a stream of numbers, and can be recorded and replayed in this form to give a totally reproducible and high-fidelity audio signal. Most commercial sound-recording studios now use digital techniques for recording.

1.3 **ELECTRONIC SYSTEMS**

It used to be necessary for students of electronics to understand at a basic physical level the way in which electronic systems worked. This is still true in so far as a student must have a thorough understanding of the physics of electronic devices. However, it is no longer necessary—or indeed possible—for students to understand the detailed workings of individual circuits. A microelectronic counter, for example, may be extremely complex in its actual operation. If a manufacturer has some novel circuit, he may be very reluctant to publish details of the way the device works. In the case of very complex microelectronic systems—such as microprocessors—it would be a major study to examine in detail the operation of just one such device. The technologist today is concerned with *systems*, and with the operating parameters of the individual circuit components. An individual circuit component in this context may be a complete amplifier or counter. The manufacturer will provide very complete data, and provided the user is fully aware of the device's characteristics, it is no longer necessary to understand the precise details of a device's internal functioning. For reasons outlined above, such information may be rather difficult to obtain.

PART I
ELECTRICITY

ELECTRICITY

What is electricity? To begin to answer this question, we have to look at the composition of matter. All matter is made up of atoms, but it is far from simple to describe an individual atom. Atoms are of course far too small to be seen under the most powerful microscope, and we have to infer their construction from the way they behave and from the way they affect various forms of radiation beamed at them. Because nobody quite knows just what an atom really looks like, physicists construct *models* of atoms to help them explain and predict atomic behaviour.

One of the simplest and most straightforward models of the atom was proposed by Niels Bohr, a Danish physicist, in 1913. In Bohr's model of the atom it is assumed that tiny electrons move in orbit around a central, much more massive, nucleus. The electrons are confined to orbits at fixed distances from the nucleus, each orbit corresponding to a specific amount of energy carried by the electrons in it. If an electron gains or loses the right amount of energy, it can jump to the next orbit away from the nucleus, or towards it. A typical atom, drawn according to Bohr's theory, is shown in Figure 2.1. Electrons towards the outer edge of an atom are

fig 2.1 *a model atom, according to Bohr's theory*

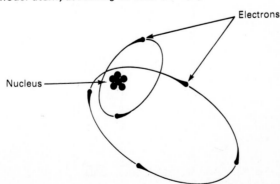

held in orbit rather loosely compared with those towards the centre. Electrons in the outermost orbits can easily be detached from the atom, either by collision or by an electric field. Once detached, they are known as *free electrons*. All materials contain some free electrons.

Materials that are classed as *insulators* contain only a very few free electrons, while materials that are classed as *conductors* (notably the metals) contain large numbers of free electrons. These free electrons drift around in the material more or less at random. However, under certain circumstances the electrons can be made to flow in a uniform direction. When this happens, they become an *electric current*.

Let us look at one such example, that of a dry cell. A study of the workings of the cell is not part of a book on electronics, but details can be found in textbooks on electricity. The illustration in Figure 2.2 shows a much simplified diagram of a typical dry cell. For our purposes, the important thing is that the internal chemistry of this cell causes the outer casing of the battery to accumulate an excess of free electrons. The centre terminal, on the other hand, has almost no free electrons. If the casing and centre electrode of the battery are connected by a conducting wire—as shown in Figure 2.3—electrons will migrate from the casing to the centre electrode by flowing down the wire. This current flow is explained by the

fig **2.2** *a single dry cell*

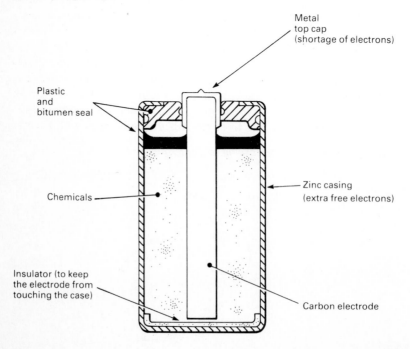

fig **2.3** *electrons flowing in a circuit*

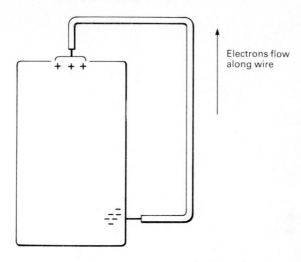

Electrons flow
along wire

fact that all electrons carry a single unit of negative electric charge. The electric charge is a form of energy, and may in some ways be regarded as analogous to magnetism. In the same way that similar poles of a magnet repel each other (north repels north, south repels south), so do electrons repel each other. In a normal atom the charges on the electrons are exactly balanced by positive charges on the nucleus. An atom in this state is said to be *neutral*. Free electrons are repelled by other free electrons, and, as you might expect, are attracted towards areas with a net positive charge. This is illustrated diagrammatically in Figure 2.4. Area *A* has a large number of free electrons, far more than are necessary to balance the positive charges on the nuclei of the atom. Area *B* has very few free electrons, and indeed a number of the atoms may be missing electrons from their outer orbits and may therefore exhibit a net positive charge. The electrons are repelled by the charge in area *A* and flow along the wire towards area *B*.

The above is something of a simplification, but illustrates the general principle.

fig **2.4** *electrons move from an area having net negative charge to an*
area with net positive charge

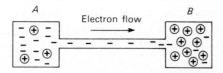

2.1 UNITS OF ELECTRICITY

It is possible to count (indirectly) the number of electrons flowing along the wire. In order for there to be a useful current, a very large number of electrons must flow. Accordingly, the basic unit of electric current flow is equivalent to around 6.28×10^{18} electrons per second. This unit of electricity is called the *ampere*, after André Marie Ampère, a French physicist who did important pioneering work on electricity and electromagnetism. The ampere, often abbreviated to 'amp', is a practical unit of current for most purposes.

In addition to measuring the current flow, we also need to measure another factor, the *potential difference* between two points. The potential difference between two points is the energy difference between them. Look at Figure 2.4 again, and imagine that there is a much larger number of free electrons on side A. This would have the effect of increasing the amount of negative charge on A and increasing the amount of energy available to push the electrons along the wire. Similarly, if side A had only a very few more free electrons than side B, the energy available to push electrons along the wire would be much less. There is a parallel here between the potential difference and water pressure—in fact we can use a 'water analogy' to illustrate the way an electric circuit works. If current flow is the equivalent of a flow of water along a pipe, the potential difference is the amount of pressure available at the end of the pipe. But whereas we might measure pressure in terms of kilograms per square metre, potential difference is measured by a unit called the *volt*. The volt is named after Alessandro Volta, an Italian physicist who, like Ampère, pioneered work in electricity—he produced the first practical battery. As you might imagine from the water analogy, the current flow is affected by the potential difference. All else being equal, the higher the potential difference—or voltage—the greater will be the current flow.

There is one other factor that would affect the flow of water in the plumbing system: that is the size of the pipe. There is a corresponding factor in the electric circuit, and that is the *resistance* of the wire. Resistance is a property of all materials that reduces the flow of electricity through them. The higher the resistance, the more difficult it is for current to flow. The three factors affecting current flow are illustrated—with the water analogy—in Figure 2.5.

The relationship between current, potential difference and resistance was first discovered by Georg Simon Ohm in 1827. It is called *Ohm's Law* after him. Ohm's Law states that the current, I, flowing through an element in a circuit is directly proportional to the voltage drop, V, across it. Ohm's Law is usually written in the form:

$$V = IR$$

fig 2.5 *a plumbing analogy: voltage, current and resistance all have their equivalents in the water system*

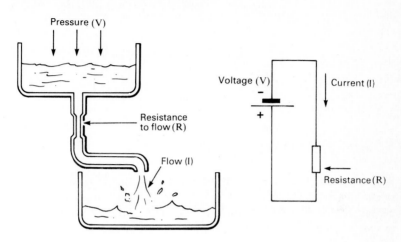

The unit of resistance is called the *ohm*, for obvious reasons. From the formula above we can see that a potential difference of one volt causes a current of one ampere to flow through a wire having a resistance of one ohm. Given any two of the factors, we can find the other one. The formula can be rearranged as:

$$I = \frac{V}{R}$$

or

$$R = \frac{V}{I}$$

This simple formula is probably used more than any other calculation by electrical and electronics engineers. Given a voltage, it is possible to arrange for a specific current to flow through a wire by including a suitable resistor in the circuit. Similarly, it is possible to determine voltage and resistance, given the other two factors. For some applications in electrical engineering, and for many in electronic engineering, the ampere and volt are rather large units. The ohm, on the other hand, is a rather small unit of resistance. It is normal for these three units to be used in conjunction with the usual SI prefixes to make multiples and submultiples of the basic units. These are given in the chart in Figure 2.6. When using Ohm's Law in calculations, it is vital that you remember to work in the right units. You cannot mix volts, ohms and milliamps!

fig 2.6 *SI prefixes*

Prefix	Symbol	Meaning	Pronunciation
tera	*T*	$\times 10^{12}$	tare-ah
giga	*G*	$\times 10^9$	guy-ger
mega	*M*	$\times 1\,000\,000$	megger
kilo	*k*	$\times 1\,000$	keel-oh
milli	*m*	$\div 1\,000$	milly
micro	μ	$\div 1\,000\,000$	—
nano	*n*	$\div 10^9$	nar-no
pico	*p*	$\div 10^{12}$	peeko
femto	*f*	$\div 10^{15}$	femm-toe

2.2 DIRECT AND ALTERNATING CURRENT

So far we have been considering electric current that flows continuously in one direction through a circuit. However, in many applications we will encounter electric current that flows first in one direction and then in the reverse direction. Current that flows first one way and then the other is called *alternating current*, and a voltage source supplying such a current is called an *alternating voltage*. Alternating voltages and currents occur in various circuits, particularly in radio and audio, and in power supplies. Alternating current is used for all mains power supplies because a.c. permits the use of *transformers* (see p. 34) to increase and decrease the voltage, so that the power can be transmitted from place to place using efficient high-voltage transmission lines.

The rate at which the voltage reverses is the *frequency* of the alternation. One complete *cycle*, from one direction to the other, and back to the original direction, is the measure: one cycle per second is given a unit, the *hertz* (symbol Hz) after Heinrich Rudolf Hertz who first produced and detected radio waves. The hertz is equal to one cycle per second, and can be used with the usual SI prefixes; thus 1000 Hz is usually written as 1 kHz, etc.

2.3 RESISTORS IN SERIES AND PARALLEL

An electrical circuit that draws power from a source of electricity is known as a *load*. It is possible to connect two loads in an electrical circuit so that all the current flows through both of them. Equally, it is possible to connect them so that current flow is divided between them. The two possibilities are illustrated in Figure 2.7.

Figure 2.7a illustrates *series* connection. In this illustration the loads have a resistance of 10 Ω and 5 Ω respectively. To work out the combined resistance of the two loads, for use in calculations involving Ohm's Law, the two values can simply be added together. This is true of any number

fig 2.7   *resistors in series (a) and parallel (b)*

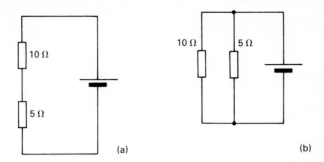

(a)   (b)

of resistive loads connected in series—just add the values together, making sure that all the resistances are expressed in the same units (once again, you cannot mix ohms and kilohms).

The second case, illustrated in Figure 2.7b, is less easy to calculate. The formula for working out combined parallel resistance is:

$$\frac{1}{R} = \frac{1}{R_1} + \frac{1}{R_2} + \frac{1}{R_3} + \dots$$

R is the combined resistance, where R_1, R_2, etc. are the individual parallel resistances. Working this out numerically for the example in Figure 2.7b, we get:

$$\frac{1}{R} = \frac{1}{10} + \frac{1}{5}$$

$$\frac{1}{R} = 0.1 + 0.2$$

$$\frac{1}{R} = 0.3$$

$$R = 3.333 \ \Omega$$

Thus the combined resistance is around 3.3 Ω. It is important at this stage to consider what degree of accuracy is required. My calculator shows that the reciprocal of 0.3 is actually 3.33333333. This seems to be a very high degree of accuracy, but the accuracy is apparent rather than actual. We must consider how accurately the true values of the two resistive loads are known. In order for the calculator's answer to be realistic, we would need to know the resistance of the load to a far greater accuracy than could be measured without the most specialised equipment. We should also consider how accurately we *need* to know the answer. For most purposes—and bearing in mind the very large number of variables that usually exist within any electric or electronic circuit—it is quite sufficient to say that the combined resistance is equal to approximately 3.3 Ω.

In practice, quite complex combinations of series and parallel resistances may be met with. Figure 2.8 illustrates such an instance. The rule for dealing with part of a circuit like this, and arriving at the combined resistance of all the loads, is to work out any obvious parallel or series combinations first, and progressively simplify the circuit. There is an obvious parallel combination in this figure, so we can work out the combined resistance of the two 100 Ω and two 50 Ω resistors first. Using a calculation like the one above, we arrive at a total resistance of about 16.7 Ω for the parallel combination of the four resistors. Figure 2.8 has now become more simple—the simplified version is illustrated in Figure 2.9a.

fig 2.8 *a complex series-parallel network*

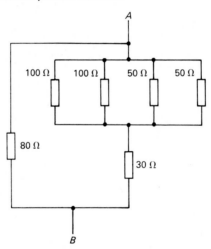

fig 2.9 *progressive simplifications of figure 2.8*

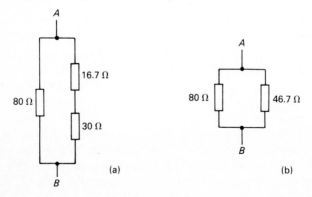

The series resistors, 16.7 Ω and 30 Ω, can simply be added together to give a total of 46.7 Ω. This leads us to Figure 2.9b, which is a simple parallel combination which can easily be worked out.

The total resistance between points A and B in Figure 2.8 is therefore 29.5 Ω—or, in the real world, about 30 Ω.

Change some of the values in the example given in Figure 2.8 and work through the calculations yourself, to make sure that you are completely happy with these simple calculations. These simple calculations will in fact be a lot simpler if you use a pocket calculator—preferably one that can give you reciprocals. Calculators are inexpensive and any student of electronics should always keep one handy.

2.4 TOLERANCES AND PREFERRED VALUES

It should be clear that a load having a resistance of, say, 20 Ω is unlikely to have a value of *exactly* 20 Ω if you measure it accurately enough. Components having a specified resistance, such as *resistors* used in electronic circuits, may be marked with a specific value of resistance. But inevitably the components will vary slightly and not all resistors stamped '20 Ω' will necessarily have a resistance of exactly this value. Manufacturing tolerances may be quite wide.

When designing circuits, it is important to bear in mind that components may not be exactly what they seem. As we saw above, this affects the calculations we make, and in any complex circuit we would have to specify circuit values of resistance, voltage and current in terms of a range of values. A resistor marked 20 Ω might, for example, have a resistance of 18 Ω, or 22 Ω. These two limits represent a departure from the marked value of around ± 10 per cent. This is in fact a typical manufacturing tolerance for resistors.

But let us look at the effect of two such resistors connected in a series circuit. Figure 2.10 shows this. We would expect, using simple addition,

fig 2.10 *the effect of tolerance in values*

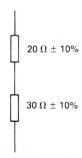

20 Ω ± 10%

30 Ω ± 10%

that the combined resistance of these two resistors would be 50 Ω. We might base our circuit design on this, and might, for example, predict that this very simple 'circuit' would draw 100 mA from a 5 volt supply. However, let us consider two extremes. If *both* resistors happen to be at the limits of their tolerance range, then the combined resistance might be 55 Ω. Our circuit would be drawing only 90 mA. If the resistors happened to be at the low end of the tolerance range, then a combined value of 46 Ω could give us a 108 mA current drain from our 5 volt supply. According to the function of the circuit, this might or might not be important.

In some cases we need to be able to specify a resistor more accurately than ± 10 per cent. Manufacturers of these components therefore produce different ranges, according to the degree of tolerance permitted. The closer the tolerance, the more expensive the resistor, so it is commercially unwise to specify component accuracy greater than the circuit requires. Because of the manufacturing tolerance, and because of the impossibility of producing *all* possible values of resistor, manufacturers restrict themselves to an internationally standardised range of values. Figure 2.11 lists the available values. Intermediate values are not generally available, but can be made by combining preferred values. It is important to be cautious when doing this, and to bear in mind the tolerance range of the resistors you are using. There is little point, for example, in adding a 3.3 Ω resistor

fig 2.11 *preferred values of resistors*

E12 series	E24 series	
10	10	*Preferred values of resistance*
	11	E12 series is available in all types of resistor
12	12	
	13	
15	15	E24 series is available in close-tolerance or high-stability only
	16	
18	18	
	20	All values are obtainable in multiples or submultiples of
22	22	10, e.g. 2.2 Ω, 22 Ω, 220 Ω, 2.2 kΩ, 22 kΩ, 220 kΩ,
	24	2.2 MΩ, 22 MΩ
27	27	
	30	
33	33	Resistance values less than 10 Ω and more than 10 MΩ
	36	are uncommon, and may not be available in all types of resistor
39	39	
	43	
47	47	
	51	
56	56	
	62	
68	68	
	75	
82	82	
	91	

to a $47\,\Omega$ resistor in an attempt to produce something that is exactly $50\,\Omega$ if the tolerance of the combination is likely to be $\pm\,5\,\Omega$. Manufacturers in general restrict themselves to three or four ranges of tolerances. Typically, small resistors would have a tolerance of $\pm$ 5 per cent, or in some cases $\pm$ 10 per cent. Close-tolerance components of $\pm\,2\frac{1}{2}$ per cent or even $\pm$ 1 per cent are available, but expensive. Components made to this accuracy would normally be used only in accurate measuring equipment or in particularly critical parts of some circuits.

2.5 CIRCUIT DIAGRAMS

I have already used circuit diagrams, and it is usually fairly clear how circuit diagrams are laid out. You will see many in this book, and should get a good idea of the way circuit diagrams should be drawn. It is normal to produce diagrams with vertical and horizontal lines, and as few diagonal or angled lines as possible. It is usually convenient to lay the circuit out so that the potential difference in the circuit (or voltage) varies from the top to the bottom of the page: this means arranging the diagram with one power supply line at the top and the other at the bottom in most cases. Circuit symbols are used to represent different components and different types of components. We have already seen the circuit symbol for a resistor and the circuit symbol for a battery. Other component symbols will be dealt with in context, and explanations will be given where necessary. There are rules about lines in the diagrams crossing, and these must be obeyed rigidly if the diagram is not to be misinterpreted. Figure 2.12 shows some of the conventions. Figure 2.12a illustrates wires crossing, but not connecting. Figures 2.12b and 2.12c show wires that are connected—the little dot is used to indicate a connection.

Sometimes where wires are crossing but not connected the illustrator will use a 'bridge' to emphasise the fact that the wires are not connected. This is illustrated in Figure 2.12d; this particular convention is now disappearing. Figure 2.12e illustrates something which should *never* appear in circuit diagrams. In all cases where wires are crossing and connected the illustration in Figure 2.12c should be used. The reason for this is that, if the circuit diagram is printed, the point where the wires cross may 'fill in' in the printing or photocopying process and make it appear that they should be connected when in fact the circuit designer intended that they should not. As you work through this book you should look at the way the circuit diagrams are laid out, and try to draw neat, *clear* diagrams yourself. Often, this may mean redrawing—but it is worth the trouble.

In the next chapter we will be looking at further 'passive' components. We have already considered resistors, but we shall look at them in more

fig 2.12 *circuit diagrams*

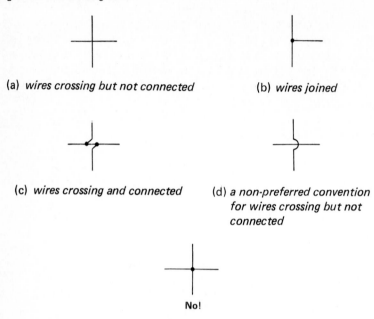

(a) *wires crossing but not connected*

(b) *wires joined*

(c) *wires crossing and connected*

(d) *a non-preferred convention for wires crossing but not connected*

No!

(e) *something that should not appear on any circuit diagram*

detail—and in the context of the way they are used in circuits. We shall also look at the other two main classes of passive components, *capacitors* and *inductors*.

QUESTIONS FOR CHAPTERS 1 AND 2

1. Describe the difference between *linear* and *digital* electronic systems.

2. What is the unit of electric current? What is the unit of electrical potential?

3. Calculate the value of the combined value of the following resistors connected in series: $33\,k\Omega$, $18\,k\Omega$, $4.7\,k\Omega$. If a battery supplies a current of $215\,\mu A$ through the resistors when they are connected to it, what is the battery's voltage?

4. What is the combined resistance of the following resistors connected in parallel: $220\,\Omega$, $100\,\Omega$, $470\,\Omega$?

5. A $10\,k\Omega$ and a $5.6\,k\Omega$ resistor are connected in series. If both resistors have a manufacturing tolerance of ± 10 per cent, what, approximately, are the maximum and minimum values of resistance we would expect to measure across the combination?

PASSIVE COMPONENTS AND POWER SUPPLIES

The components used in electronic circuits are divided into two main classes. These are referred to as 'active' and 'passive' components. Active components use the movement of electrons either through a vacuum or through a semiconductor to perform some function in the circuit. Passive devices, on the other hand, do not control the movements of the electrons, but rather allow the electrons to flow through them. It could be argued that the simplest of all passive devices is a length of wire!

3.1 RESISTORS

Resistors are components intended to insert a known amount of resistance into an electrical circuit. As we saw in the last chapter, they are available in a range of different values and tolerances. It will also be evident to anyone who has looked at an electronic circuit that they are also available in a range of *sizes*. The sizes relate to the amount of power that the resistor can dissipate. An electric current passing through a resistor will always cause a certain amount of energy to be dissipated as heat. The amount of energy dissipated can be calculated by the simple formula:

$$P = VI \text{ watts}$$

where P represents power.

Let us consider the case of a $5\,\Omega$ resistor connected across a 12 volt battery. This is illustrated in Figure 3.1. We can use Ohm's Law to calculate the amount of current flowing through the resistor. This is given as:

$$I = \frac{V}{R}$$

$$I = \frac{12}{5}$$

$$I = 2.4 \text{ A}$$

fig 3.1 *power: a simple circuit like this will dissipate heat*

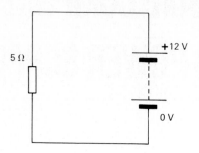

It is clear that the voltage, when measured across the resistor, must be equal to the battery voltage, provided we neglect any resistance in the wires. We can now use the formula to calculate the amount of power that the resistor is dissipating as heat:

$$P = VI$$
$$P = 12 \times 2.4$$
$$P = 28.8 \text{ watts}$$

28.8 watts is a respectable amount of heat—the above calculation might be made for a car's rear-window heating element. It is quite obvious that a tiny component such as a resistor one might find in an electronic circuit could not possibly dissipate this much heat without damaging itself. In fact, a resistor designed to dissipate around 30 watts would be quite large —perhaps 5 cm long. It would also be made in such a way that the body of the resistor could become quite hot. In practice, the designers of most electronic circuits try to avoid such large heat outputs which can cause problems in cooling the equipment.

Resistors used in electronic circuits are designed with power dissipations of 0.25, 0.5 or 1 watt. Larger sizes are available, but are less commonly used.

The same formula as the one above is used to calculate the power dissipation of series and parallel combinations. Figure 3.2 shows such circuits. Figure 3.2a illustrates two resistors connected in series. It is quite clear that the same current flows through both resistors. However, when measured across each resistor, the voltage will be different. The total potential difference is given as 5 volts. We can find the voltage across each individual resistor by using Ohm's Law. We know that the total resistance of the circuit is 3.2 kΩ and can use Ohm's Law to calculate that the current flowing through the circuit is about 1.562 mA. Using the form of Ohm's

Law $V = IR$, we can calculate the voltage drop across each resistor. For the first resistor:

$$V = 1000 \times 1.562 \, mV$$
$$V = 1.562$$

And for the other resistor:

$$V = 2200 \times 1.562$$
$$V = 3.44$$

Now that we know the voltage dropped across each resistor we can calculate the amount of power that each resistor is dissipating. For the $1 \, k\Omega$ resistor this is:

$$P = 1.562 \times 1.562$$
$$P = 2.43 \, mW$$

And for the $2.2 \, k\Omega$ resistor the power dissipated is given as:

$$P = 1.562 \times 3.44$$
$$P = 5.37 \, mW$$

Look carefully at the units used in these calculations! The amounts of power dissipated are very small, typical, in fact, of the sort of thing we will become used to in electronic circuits. The smallest commonly available size of resistor (0.125 watts) is substantially larger than is necessary to dissipate this amount of power.

Exactly the same kind of calculation is used to work out the power dissipations in resistors in parallel. In this case—illustrated in Figure 3.2b— the voltage across both resistors is the same (5 V), but the current is different. We can use Ohm's law to ascertain the current flowing through

fig 3.2 *resistors in series and in parallel*

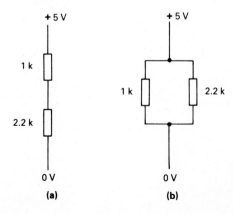

(a) (b)

each resistor, and arrive at 5 mA and 2.27 mA for the 1 kΩ and 2.2 kΩ resistors respectively. We can now use our power calculation to determine that the 1 kΩ resistor is dissipating $5 \times 5 = 25$ mW. The 2.2 kΩ resistor is dissipating $2.27 \times 5 = 7.35$ mW. If you are in any doubt about this calculation, work it out on a piece of paper. Another, equally valid way of working out power dissipations in series or parallel combinations of two resistors is to calculate the total dissipation of the pair (very simply done) and then use a ratio calculation to determine how much power is dissipated by which resistor. Note that in *series* combinations it is always the resistor with the *higher* value that dissipates the most power. In *parallel* combinations, it is the resistor with the lower resistance that dissipates more power.

Before leaving resistors, we should look at the physical structure of the three main types of resistor. These are all electrically comparable—there is only one kind of resistance—but they are mechanically different, being suitable for slightly different applications.

Types of resistor

The most usual type of resistor is the solid carbon resistor (see Figure 3.3). Its structure is very simple, consisting of a small cylinder of carbon which is mixed with a non-conductor. A connecting wire is fixed into each end, and the resistor is given a coat of paint to protect it from moisture which might alter the resistance.

Resistors are always marked with a *colour code* to indicate the value. The colour code consists of three or four coloured bands painted round the resistor body. This system is used because it makes the resistor's value

fig **3.3**   *a solid carbon resistor (this component might be anything from a few millimetres to a few tens of millimetres long)*

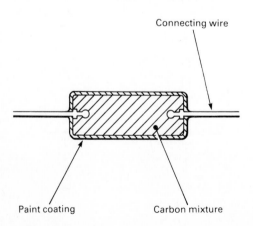

Connecting wire

Paint coating   Carbon mixture

visible from any direction—a printed label could be hard to read with the component in place on a crowded board. Also, a painted or printed value could easily get rubbed off, whereas painted bands are relatively permanent.

The first three bands of the colour code represent the value of the resistor in ohms. Bands 1 and 2 are the two digits of the value, and band 3 represents the number of zeros following the first two digits. Figure 3.4 gives the resistor colour code; the standard is international. The fourth band is used to indicate the tolerance of the resistor's stated value.

fig **3.4** *the international standard resistor colour code*

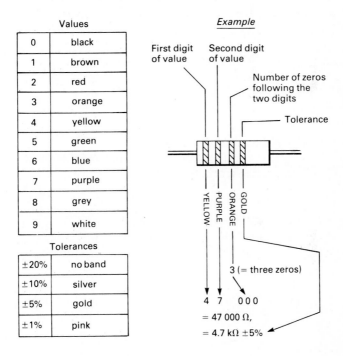

Where values of resistance less than two digits are required, the two bands are followed by a gold band. Thus 4.7 Ω would be represented as yellow, purple, gold, plus a tolerance band.

The next kind of resistor is the *metal oxide* or *metal glaze* resistor. This looks rather like the carbon resistor from the outside, but the internal structure is different. Figure 3.5a shows a cross-section.

Metal oxide resistors can be made to closer tolerances than carbon resistors, and change their resistance less with changes in temperature. For this reason they are sometimes called *high-stability* resistors. The resistance of a metal oxide resistor changes by approximately 250 parts per million

fig 3.5

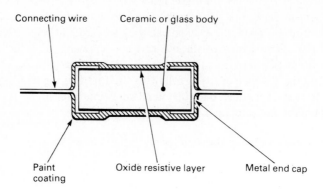

Connecting wire Ceramic or glass body

Paint coating Oxide resistive layer Metal end cap

(a) *a metal oxide resistor*

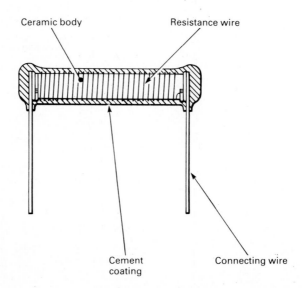

Ceramic body Resistance wire

Cement coating Connecting wire

(b) **for high power applications, a wire-wound resistor is used**

per °C. This compares with about 1200 parts per million per °C for carbon resistors.

Both carbon and metal oxide resistors are made in a range of stock sizes, from about 0.125 W dissipation to about 3 W. Sometimes it is necessary to have resistors that can cope with higher powers; for this, *wire-wound* resistors are used. The wire-wound resistor is shown in Figure 3.5b.

It is easy to make wire-wound resistors in low resistance values, down to fractions of an ohm. High resistance values use wire of low conductivity.

requiring many turns of fine gauge wire as well. The maximum practical value for a wire-wound resistor is a few tens of kilohms, at least for components that are a reasonable size. Power ratings range from 1–50 W in the stock sizes; there is no limit to the size in practice, and larger special-purpose wire-wound resistors are common.

It is possible to make *precision wire-wound resistors* in which the resistance is specified to a very close tolerance, within ± 0.1 per cent. Such resistors are expensive and would be used only in measuring equipment.

Variable resistors
For applications such as volume controls and other 'user controls' in electronic equipment, it is often necessary to have a resistor that can be altered in resistance by means of a control knob. Such resistors are called *variable resistors*, or *potentiometers*. They consist of a resistive *track*, made with a connection at either end. A movable *brush*, generally made of a non-corroding metal, can be moved along the track; an electrical connection to the brush allows a variable resistance to be obtained between either end of the track and the brush.

Three forms of variable resistor are shown in Figure 3.6b. Note that the shaft (or slot) that moves the brush is generally connected to the brush; it is important to know this for safety reasons.

Variable resistors are available in various shapes and sizes, with power dissipations from around 0.25 W upwards. Tracks are either carbon, conductive ceramic ('cermet') or wire-wound. Resistance ranges are available between fractions of an ohm and a few megohms.

The track can be made in such a way that the resistance increases smoothly along the track—the usual, *linear*, type of potentiometer. For some audio uses (such as volume controls) *logarithmic* potentiometers are made in which the resistance increases according to a logarithmic law rather than a linear law. For most purposes you need only remember that the way logarithmic potentiometers control volume approximates the way the human ear responds to sounds of different loudness. If you use a linear potentiometer for a volume control, the effect of the control seems to be 'all at one end' of the scale.

3.2 CAPACITORS

Next to resistors, the most commonly encountered component is the *capacitor*. A capacitor is a component that can store electric charge. In essence, it consists of two flat parallel plates, very close to each other, but separated by an insulator (see Figure 3.7).

When the capacitor is connected to a voltage supply, a current will flow through the circuit (see Figure 3.8). Electrons are stored in one of the

fig 3.6

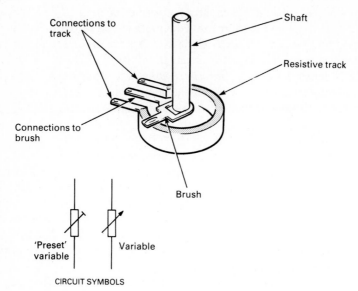

CIRCUIT SYMBOLS

(a)   *principle of the potentiometer or variable resistor*

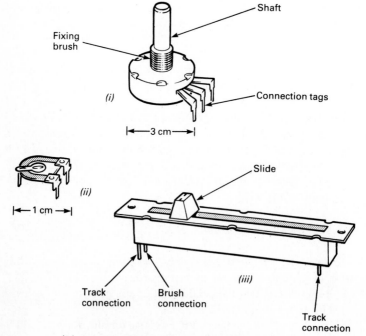

(b)   *three different types of potentiometer*

fig 3.7 *schematic view of a capacitor*

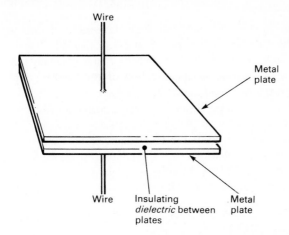

fig 3.8 *a capacitor connected to a power supply; current will flow until the capacitor is charged*

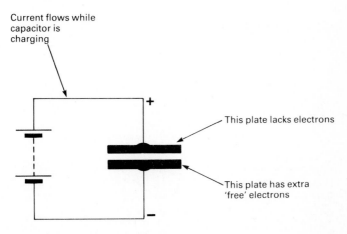

plates of the capacitor; in the other there is a shortage of electrons. In this state the capacitor is said to be *charged*, and if it is disconnected from the supply, the imbalance in potential between the two plates will remain.

If the charged capacitor is connected in a circuit, it will, for a short time, act as a voltage source, just like a battery. This can be demonstrated quite nicely with nothing more than a large capacitor, a 9 V battery such as PP9, and a capacitor rated at 10 V, 1000 µF (see below regarding *units of capacitance*). Note that the capacitor will be an *electrolytic* type (again, see below) and must be connected to the battery with its + terminal towards the positive (+) terminal of the battery. The demonstration is shown in Figure 3.9.

Try the experiment again, this time allowing two minutes between charging the capacitor and discharging it into the bulb. Try waiting progressively longer, and you will see that the charge gradually leaks away on its own. This is due to the imperfection of the insulator separating the plates, allowing a tiny *leakage current* to flow.

A graph of the current flowing through a charging and then discharging capacitor is compared with the voltage measured across it in Figure 3.10.

The unit of capacitance is the *farad*, named after Faraday. One farad is a very large unit of capacitance, in the context of electronic circuits, and the smaller derived units of capacitance are used. The largest practical unit is the *microfarad*, symbol μF. Although values of capacitance in the range

fig 3.9 *a simple experiment for demonstrating the way in which capacitors can store power; a large capacitor is first charged up from a 9 volt supply, and then discharged into a small lamp*

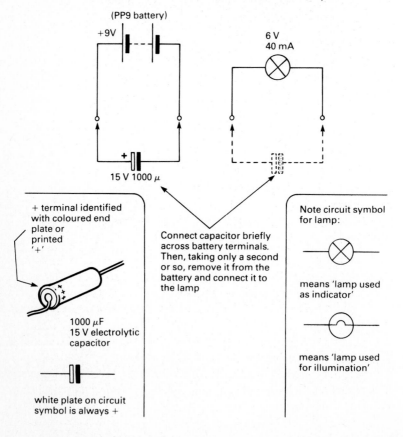

(PP9 battery)

+9V

6 V
40 mA

15 V 1000 μ

+ terminal identified with coloured end plate or printed '+'

Connect capacitor briefly across battery terminals. Then, taking only a second or so, remove it from the battery and connect it to the lamp

Note circuit symbol for lamp:

means 'lamp used as indicator'

1000 μF
15 V electrolytic capacitor

means 'lamp used for illumination'

white plate on circuit symbol is always +

fig 3.10 *a graph showing a capacitor charging and discharging*

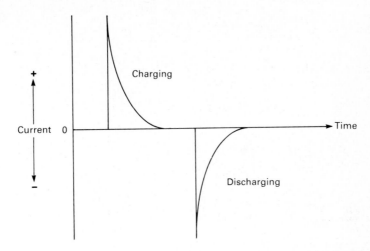

of thousands of microfarad are sometimes used, the millifarad is never encountered—a capacitor would be marked 10 000 μF, for example, not 10 mF. If you do see a capacitor marked 'mF', it is almost certainly meant to be μF, with that manufacturer (Far Eastern?) using a non-standard abbreviation. In addition to the μF, the nF and the pF are commonly used units.

There are many different types of capacitors, according to the use to which the component is to be put, and also to the operating conditions. The capacitance of the component is determined by three factors: the area of the plates, the separation of the plates, and the insulating material that separates them, known as the *dielectric*. For the same dielectric material, the closer and larger the plates, the greater the capacitance. Another factor determines the thickness of the dielectric, and that is the maximum voltage to which it is going to be subjected. If the dielectric is very thin, it will *break down* with a relatively low voltage applied to the plates of the capacitor. Once the dielectric has been damaged, the capacitor is useless. High voltage capacitors need thicker dielectrics, to withstand the higher voltage. To produce the same capacitance, the plates have to be larger in area—so the component is bigger.

The simplest type of capacitor uses a roll of very thin aluminium foil, interleaved with a very thin plastic dielectric such as *mylar*. A physically smaller capacitor can be made by actually plating the aluminium on to one side of the mylar (see Figure 3.11). For higher voltages, *polyester, polystyrene* or *polycarbonate* plastic material is used as the dielectric.

Ceramic capacitors are used where small values of capacitance and large values of leakage current are acceptable—the ceramic capacitor is inexpen-

fig 3.11 *a typical capacitor, used in electronics work—the 'metallised film' type*

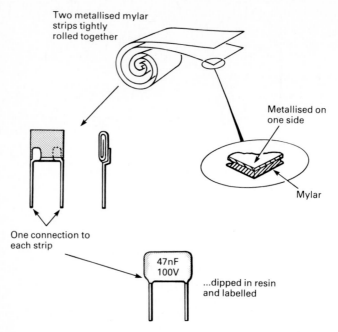

Two metallised mylar
strips tightly
rolled together

Metallised on
one side

Mylar

One connection to
each strip

47nF
100V

...dipped in resin
and labelled

sive. A thin ceramic dielectric is metallised on each side, and coated with a thick protective layer, usually applied by dipping the component.

Both plastic film and ceramic capacitors are available in the range $10\,pF$-$1\,\mu F$, though plastic film types may be obtained in larger values.

Where really high values of capacitance are needed, *electrolytic capacitors* are used. Electrolytic capacitors give very large values of capacitance in a small component, at the expense of a wide tolerance in the marked value (-25 to $+50$ per cent) and the necessity for connecting the capacitor so that one terminal is always positive.

The most commonly used type of electrolytic capacitor is the *aluminium* electrolytic capacitor. The construction is given in Figure 3.12. After manufacture, the capacitor is connected to a controlled current source which electrochemically deposits a layer of aluminium oxide on the surface of the 'positive' plate. The aluminium oxide makes an excellent dielectric, with very good dielectric strength (resistance to voltage applied across the plates). Since the layer is chemically deposited it is very thin.

The electrolytic capacitor must not be subjected to voltages applied in the wrong direction, or the aluminium oxide layer will be moved off the positive plate, and back into the electrolyte again.

fig **3.12** *construction of an aluminium electrolytic capacitor*

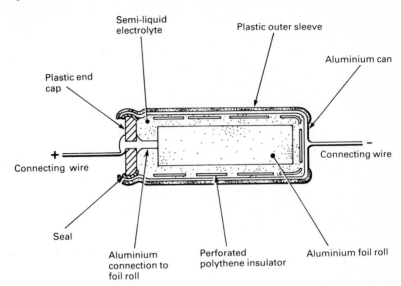

Variable capacitors

Capacitors are made that can be varied in value. Either air or thin mica sheets are used as the dielectric, so the capacitance of variable capacitors is usually low. *Rotary* or *compression* types are made. In the former, the capacitance is altered by changing the amount of overlap of the two sets of plates; see Figure 3.13a. In compression trimmers, the spacing between plates is altered, as in Figure 3.13b.

Variable capacitors can be obtained with maximum values from $2\,pF$ to $500\,pF$.

Capacitors as frequency-sensitive resistors

Once a capacitor has charged up, it will not pass current when connected to a direct voltage supply. However, if we *reverse* the polarity of the supply, the capacitor will permit a current to flow until it has charged up again, with the opposite plates positive and negative. The capacitor will then, once again, block the flow of current. If the frequency of an alternating voltage applied to a capacitor is high enough, the capacitor will actually behave as if it were a low value resistor, as it will never become charged in either direction. At lower frequencies, the capacitor will appear to have a higher resistance, and at zero frequency (d.c.) it will, as we have seen, have an infinitely high resistance if you disregard the leakage current.

The precise value of resistance shown by the capacitor will depend upon the capacitance, the applied voltage and the frequency. The property

fig 3.13

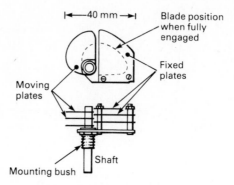

40 mm

Blade position
when fully
engaged

Fixed
plates

Moving
plates

Mounting bush

Shaft

(a) *a variable capacitor of the 'air space' type (this component would be used for tuning a radio receiver, for example)*

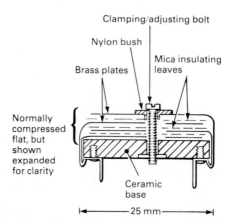

Clamping/adjusting bolt

Nylon bush

Mica insulating
leaves

Brass plates

Normally
compressed
flat, but
shown
expanded
for clarity

Ceramic
base

25 mm

(b) *a 'compression trimmer', a form of variable capacitor used to preset circuit values (adjustment is by means of a screw)*

is known as *reactance*. It enables the circuit designer to use the capacitor to block or accept different frequencies in a variety of different circuit configurations, many of which you will meet later in this book.

Capacitors in series and parallel

Capacitors can be used in series and in parallel. When used in series, the working voltage is the sum of the two working voltages, so two capacitors intended for a 10 V maximum supply voltage could be used, in series, with a 20 V supply.

The calculation of capacitor values in series and parallel is similar to the calculation used for resistors in series and parallel, but in the opposite case. Thus for capacitors in *parallel* the formula:

$$C = C_1 + C_2 + C_3 + \ldots$$

is used, simply adding the values together. For capacitors in *series*:

$$\frac{1}{C} = \frac{1}{C_1} + \frac{1}{C_2} + \frac{1}{C_3} + \ldots$$

gives the total value of capacitance.

Unfortunately, there is no internationally agreed colour code for capacitors, so manufacturers usually stamp the value on a capacitor, using figures in the normal way. A few capacitors are marked with coloured bands, but it is wise to look at the maker's catalogue to check what the code means.

3.3 INDUCTORS

The third major passive component is the inductor. This is a coil of insulated wire that may, or may not, be wound over a ferrous metal former. When a coil is connected in a circuit, as in Figure 3.14, the flow of current through the coil causes an electromagnetic field to be created around the coil. Building the field absorbs energy, and the electromagnetism produces its own current in the coil, opposing the direction of the applied current. The effect is that, when current is first applied to the inductor, it seems to have a high resistance, reducing the current flow through it. Once the electromagnetic field is established, the resistance drops. A graph of the current flowing through an inductor, and the voltage measured across it, is given in Figure 3.15.

Compare this graph with the one in Figure 3.10, and you will see that the inductor is the 'opposite' of a capacitor! Once the current flowing through the inductor is constant, the only opposition to current flow comes from the resistance of the coil wire—usually small compared with the apparent resistance (called inductive *reactance*) when the current is building up.

The opposing force that prevents current flowing right away is *induced current*, trying to flow the opposite way. We can make great use of this phenomenon in constructing a *transformer*.

The unit of inductance is the *henry*, symbol H. In electronics, mH and μH are commonly found, but the henry, like the farad, is rather a large unit.

Inductors are available as a range of components, in small inductance values from a few μH to a few mH. However, there are so many variable

fig 3.14 *an inductor connected in a simple circuit; the current rises to a fixed level, determined by the resistance of the inductor's windings*

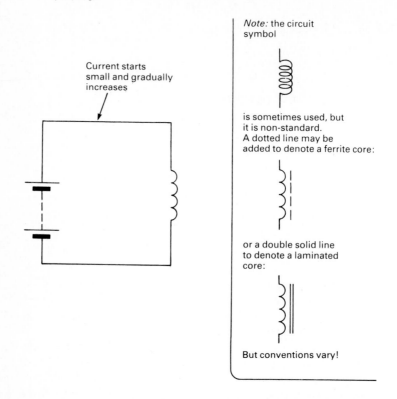

Current starts small and gradually increases

Note: the circuit symbol

is sometimes used, but it is non-standard. A dotted line may be added to denote a ferrite core:

or a double solid line to denote a laminated core:

But conventions vary!

factors that a 'stock range' of all required types would be too large for a supplier to produce. Where inductors are needed (and they are relatively rare in modern circuits) they may well be specified in terms of wire thickness, number of turns, core type, etc.

3.4 TRANSFORMERS

Transformers make use of *mutual inductance*, in which a current flowing in a coil produces an electromagnetic field, which in turn induces a current to flow in a second coil wound over the first one.

The construction of a transformer is shown in Figure 3.16. The iron laminated core is used to concentrate the electromagnetism and thus improve the efficiency. It is important that the iron laminations are insulated from each other; if they are not, the core itself will behave as if it

fig 3.15 *graphs showing voltage and current in the circuit of Figure 3.14*

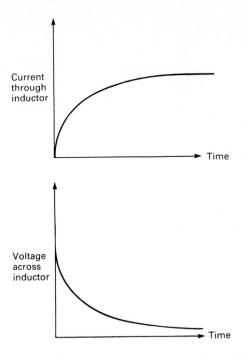

Current
through
inductor

Time

Voltage
across
inductor

Time

fig 3.16 *a typical small transformer, showing the iron laminations*

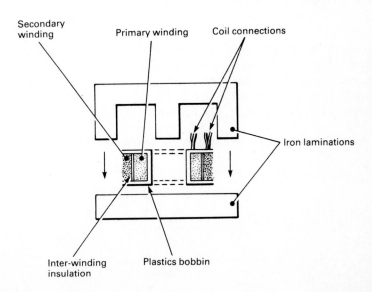

Secondary
winding

Primary winding

Coil connections

Iron laminations

Inter-winding
insulation

Plastics bobbin

were a one-turn coil, and a current will be induced in it. The current would be very large, and cause the transformer to overheat.

The ratio of the *input voltage* to the primary winding to the *output voltage* from the secondary winding depends on the turns ratio of the coils, and is approximately equal to that ratio. Thus a transformer with 2000 turns on the primary and 1000 turns on the secondary will have an output voltage that is half the input voltage.

Since currents are induced only by *changing* magnetic fields, the transformer will operate only if the input voltage is constantly changing. The transformer could be used with alternating current (such as the mains supply), and one of the main uses for a transformer in electronics is to reduce the voltage of the a.c. mains to a lower level, suitable for electronic circuits. This is illustrated in Figure 3.17.

Transformers are rated according to the turns ratio, the power handling capability (in watts), and the type of application. *Power transformers*, intended for power supplies, usually have mains voltage primaries, and a high standard of insulation between the primary and secondary coils, for safety.

The secondary winding may have a range of connections, or *taps*, so the transformer can be used in different applications. It is important to realise that the transformer simply transforms voltages; there is no net power gain, always a small loss. If a transformer has a 200 V primary and a 1000 V secondary, the voltage will be increased by a factor of five times, but it will require more than five times the current in the primary than will be available from the secondary. You can trade current for voltage and vice versa, but the *power* will always be slightly less at the output, due to various losses in the transformer (dissipated, as usual, in the form of heat).

Any component having a coil carrying current will have an inductive characteristic, though it may be swamped by other factors (resistance or capacitance). This may sometimes be important. For example, a *relay* is an electromagnetic switch, the coil being used to operate a mechanical switch. However, the coil has inductance. A relay operating from a 6 V supply can, when suddenly disconnected, produce an output pulse of

fig 3.17 *circuit symbol for a transformer with a laminated core*

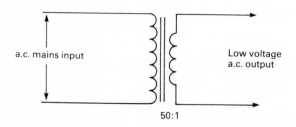

50:1

100 V or more as the current flowing through it drops very rapidly to zero. This voltage may be high enough to harm semiconductor components, and precautions should be taken to ensure that this 'spike' of high voltage is made harmless.

3.5 POWER SUPPLIES

I am going to introduce, prematurely, a semiconductor component that is described fully in Chapter 7. This is the *diode*, a sort of electronic one-way street that will let current pass through it in one direction but not in the other direction.

Diodes are used in power supply circuits, in association with transformers. A transformer is used to reduce the voltage of the a.c. mains to something more usable in electronic circuits, say 10 V in this case. However, the output is still a.c., and most electronic circuits require a direct current supply. The transformer's output can be converted, or *rectified*, by one or more diodes. Figure 3.18 shows a transformer operating a circuit with a bulb. The waveform of the voltage across the bulb is shown also.

Now compare the waveform with the one in Figure 3.19. The diode has blocked current flowing in one direction, so the top connection of the bulb is always positive. Only half the power reaches the bulb, so this is not a very efficient design. If four diodes are used, in a configuration called a *bridge rectifier*, the whole of the current available from the transformer can be passed through the bulb. Follow the current path through Figure 3.20, first for one direction of output from the secondary, and then from the other.

The supply to the bulb is now d.c., but constantly changing in voltage. The voltage supplied to the bulb (or any other circuit) can be *smoothed* by means of a large-value capacitor connected in parallel with the load (the bulb in this case). The effect of the capacitor is to supply the required

fig **3.18** *waveforms at the input and output of a transformer driving a lamp*

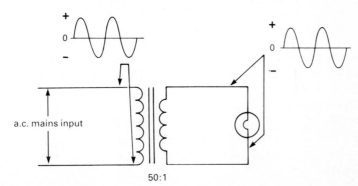

a.c. mains input

50:1

fig 3.19 *waveforms for half-wave rectified direct current, supplied by the secondary of the transformer, through a diode*

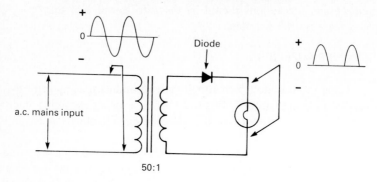

fig 3.20 *the diode 'bridge rectifier' configuration, used to provide full-wave rectification of an a.c. supply (although both halves of the a.c. input waveform pass through the lamp, the current flow through the lamp is not smooth—and this would upset the operation of many circuits)*

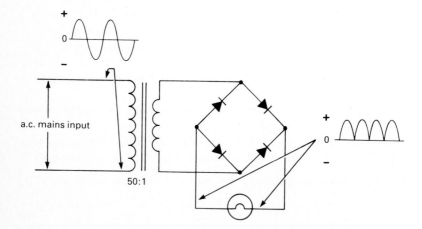

current when the voltage drops during each half-cycle of the mains supply. When the voltage rises, the capacitor is recharged, so that it has energy available to fill in the 'gaps' in the supply. Figure 3.21 illustrates this.

This circuit is the basis of almost all power supply systems used in electronic circuits. Although much more sophisticated 'regulated' power supplies are often used—as you will discover when you get to Chapter 20—the transformer, bridge rectifier and smoothing capacitor feature in all the designs.

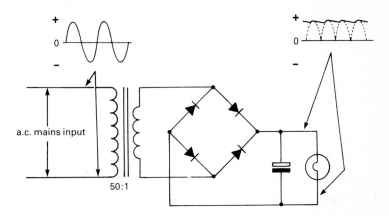

fig **3.21** *the bridge rectifier with smoothing capacitor; a large capacitor is used to provide power during the 'gaps' in the rectified waveform*

Before leaving the subject of diodes and circuits, I will use this opportunity to mention an important fact that electronics engineers, secretly, are a little ashamed of.

The current flowing through the power supply circuits above is flowing in the direction of the arrowhead of the diode symbol. The arrowhead to indicate current flow occurs over and over again in circuit diagrams. Current, we say, always flows from positive to negative.

Actually, it doesn't.

Electrons move the other way, from negative to positive, opposite to what has been called 'conventional current'. This is an odd fact, but one that only rarely complicates circuit design and analysis. The reason for it, frankly, is a Mistake. When the study of electricity was in its infancy, there was a need to standardise (in books and notes) on the direction of current flow. In the absence of our knowledge of electrons and atoms, someone sat down to perform an experiment that would settle the problem of current flow once and for all. It might even have been Volta, or Faraday, I can't remember. The experiment consisted of setting up equipment that would make a very long electric spark, and then repeatedly watching the spark. Eventually the experimenter made up his mind that the spark always jumped from *here* to *here*. It was at this point that the Mistake was made.

When, years later, it was discovered that (i) electrons go from negative to positive, and (ii) electric sparks jump too fast to see which way they are going anyway, 'conventional current' was established in all the textbooks, and it would have been disastrous to have tried to change it.

So we *still* say that 'conventional current' flows from positive to negative, except when we are discussing semiconductor physics, where electron flow is what matters.

3.6 PRINTED CIRCUITS

Before leaving passive components, mention must be made of that most ubiquitous of passive components, the *printed circuit board*. It used to be necessary to build a *chassis* for every piece of electronic circuitry, with insulating strips carrying rows of tags for component connections. The printed circuit board (PCB), however, has made possible small, light-weight circuits, easy to construct and repair.

The PCB consists of a strong plastic laminate or fibreglass laminate board, with a thin sheet of copper bonded to one (or sometimes both) side(s). The sheet is drilled with holes that enable the circuit components to be pushed into place with their connecting leads through the holes. The underneath (copper) side of the PCB is made in a pattern which corresponds to the required wiring pattern. The components are soldered in place, and the PCB provides both support for the parts and the necessary electrical interconnections. A completed PCB can be seen in Figure 15.23 (p. 219). Appendix 1 shows some examples of PCB patterns.

In production, the PCB is produced using automatic processes. The required pattern is produced as a photographic negative, and this is printed on to the PCB, which has been coated on the copper side with a photo-graphic emulsion. The PCB is then developed like a photograph, leaving emulsion only on the parts that were exposed to light.

Next, the PCB goes into a bath of a powerful chemical that will dissolve copper—ferric chloride is often used. The ferric chloride etches away all the copper, *except* where it is protected by the emulsion. The emulsion is cleaned off, leaving the copper pattern. Holes are drilled or punched as required, and the board is ready for assembly (although the copper is often coated with a thin layer of solder by a process known as *roller tinning* to improve ease of soldering.

Anyone can make PCBs. For single PCBs, the circuit is simply drawn on to the copper with an etch-resistant ink (special pens are available) or with rub-on transfers. The rest of the process is the same as that used com-mercially.

Be careful with the strong ferric chloride solution—it is corrosive, poisonous, and will gallop through most metals. Use rubber gloves, plastic dishes, plastic tongs, and *don't* work on a stainless steel drainer!

* * * * * * * * * * * * * * * * * *

Electricity and Passive Components

Chapters 2 and 3 have been intended as revision only. If you want to know more about electricity, there are many good textbooks available. If, however, you have found the first three chapters of this book fairly easy going, you will be able to understand the rest without too much trouble.

QUESTIONS

1. State the value and tolerance of the following resistor colour codes, as read from the resistors in a circuit board in a radio receiver: *brown, black, red, silver*; *brown, grey, yellow, silver*; *red, black, black, silver*; *gold, red, blue, green.*

2. What is the largest commonly used unit of capacitance?

3. Three capacitors are connected in parallel. Their values are $4.7\,\mu F$, $10\,\mu F$ and $3.3\,\mu F$. What is the combined capacitance? If the tolerance of each is $+20$ per cent and -10 per cent, what are the maximum and minimum values of capacitance to be expected for the combination?

4. The following two capacitors are connected in series: $3.3\,nF$, $2.2\,nF$. What is the combined capacitance?

5. A transformer is taking a current of $500\,mA$ from a $220\,V$ a.c. supply. If the output of the transformer is at $440\,V$, what is the maximum current the transformer could deliver from its secondary winding?

CHAPTER 4

TOOLS, TEST EQUIPMENT AND SAFETY

4.1 TOOLS

Soldering

Components are connected together in circuits by soldered joints. Solder is basically an alloy of lead and tin, the proportions varying according to the application. For electronics work a mixture of 60 per cent tin and 40 per cent lead is usual. If solder is heated to its melting point, and applied to a variety of different metals, it will amalgamate with the metal surface to provide a joint with high electrical conductivity and good mechanical strength—the strength is limited by the relatively poor tensile strength of the solder itself. Solder has a low melting point, lower than that of either tin or lead. The melting point varies according to the exact composition of the solder, but ordinary 60/40 tin/lead solder melts at 188°C.

The temperature required to melt the solder is also sufficient to oxidise many metals, forming a layer of oxide that will prevent the solder 'sticking' to the surface. Solder intended for electrical work is therefore made in the form of a hollow wire, with one or more cores of resin flux, a chemical mixture that will dissolve the oxide film at soldering temperatures. A cross-section through two different makes of electrical flux-cored solder is given in Figure 4.1. Solder can also be obtained in bar form, without the flux core.

For prototype and repair work, soldering is carried out using a *soldering iron* or *soldering gun*. The *bit* or tip of the iron is heated electrically to between 350 and 420°C. To make the joint, the bit of the iron is applied to the two surfaces to be joined, and the wire solder applied to the heated joint. The solder melts, allowing the flux to run over all surfaces and clean them. The solder also helps conduct heat from the iron on to the surfaces, and when the temperature is high enough it amalgamates with them. The iron can now be removed and the finished joint allowed to cool.

The means of heat generation is different in soldering irons and solder-

fig **4.1** *two types of resin-cored solder*

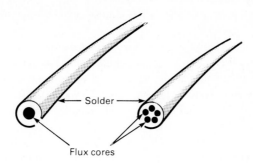

fig **4.2** *a soldering gun—ideal for 'heavy' soldering and some kinds of repair work*

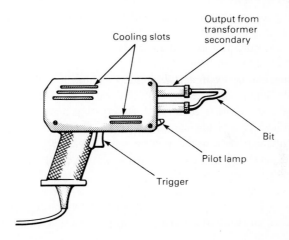

ing guns. A soldering gun is illustrated in Figure 4.2. The gun is basically a transformer. The secondary winding has a few turns of very heavy copper wire or bar, and generates a very heavy current at low voltage. The bit of the gun is made of copper, and, being substantially thinner than the transformer secondary windings, has a relatively high resistance. The energy generated by the transformer secondary winding heats the copper bit rapidly, until the soldering temperature is reached.

Soldering guns have the advantage that they are quick to heat up—only a few seconds—and cool down rapidly after switching off. The gun is on only when the trigger is held down. Soldering guns are also powerful, and are useful for soldering large components or small metal sheets. A typical soldering gun produces a heat output of some 50–150 watts.

The main disadvantage is that heat regulation is not particularly good, and it needs a certain amount of skill to keep the bit at the proper temperature. Also, the bit is large and rather clumsy, making the gun unsuitable for fine work.

Soldering irons are used for most work. A soldering iron uses a heating element to bring the bit up to the required temperatures. A small iron for electrical work would have an output of 7–25 watts, depending on size; for delicate work (e.g. integrated circuits) about 10 watts is sufficient.

Some types of integrated circuit are damaged by the slightest electrical leakage of high voltage from the mains supply, through the insulation of the iron (or through internal capacitance). To reduce this danger, *ceramic shafted* soldering irons are available, in which a thin ceramic sleeve insulates the bit from the rest of the iron. The sleeve has excellent insulating properties and low capacitance, and provides a good measure of safety when soldering sensitive circuits. A sectional drawing of a ceramic-shafted solder-iron is given in Figure 4.3.

Soldering is something of an art, and needs to be practised. Typical bad joints are caused by insufficient heat, dirty surfaces, insufficient solder or persistent reheating of a joint in an attempt to get it to stick. A selection of problems is shown in Figure 4.4.

Some metals—for example, aluminium—cannot be soldered by normal techniques. Metals which can be soldered easily include gold, silver, tin, copper and lead. Gold is often used as very thin plating over copper or iron component connecting wires, to improve solderability.

In commercial batch production lines *wave soldering* is often used. In this system, all the components are first placed in position in the printed circuit board, and the leads cut off to the right length. The whole circuit board is then moved across a special soldering bath, to solder all the component connections at once. The soldering bath consists of a bath of molten solder, pumped over a baffle in such a way that a standing wave is

fig 4.3 *a ceramic shafted soldering iron, used for light soldering jobs, especially where circuits that are sensitive to electrostatic voltages are being soldered*

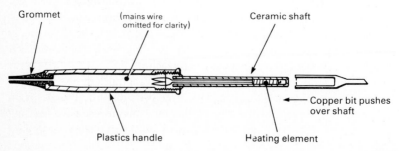

Grommet

(mains wire omitted for clarity)

Ceramic shaft

Copper bit pushes over shaft

Plastics handle

Heating element

fig 4.4 *some soldered joints, in cross-section*

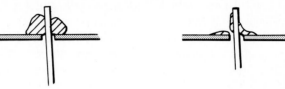

(a) *soldered joint too cold*

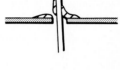

(b) *insufficient solder*

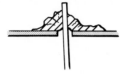

(c) *solder reheated too often*

(d) *wire not far enough into hole; this results in a joint that is mechanically weak*

(e) *too much solder forms a 'bridge' to the next part of the circuit board*

(f) *a 'dry joint': the solder has stuck to the board, but not to the wire, perhaps because the wire was dirty; difficult to detect, and can develop an high electrical resistance*

(g) *correctly made joint*

produced, making a 'hump' in the surface of the molten solder. The bottom of the circuit board is passed over the hump, and solder sticks to the board and the components but not to the areas of plastic or varnish that are not required to be soldered. The mass of the solder is of course ample to hold sufficient heat to bring the component connections and the copper leads up to soldering temperature almost immediately.

Desoldering

Usually a component can be heated up with the soldering iron and the connections pulled apart. In some cases this is not practicable, for example where an integrated circuit is soldered directly into a printed circuit. All the pins—as many as forty—would have to be heated up simultaneously before the component could be removed from the board. To assist in removing this sort of component, a *solder sucker* is used. A solder sucker is illustrated in Figure 4.5.

fig **4.5** *a solder sucker; the nozzle is made of polytetrafluoroethylene (p.t.f.e.), a plastic that is resistant to heat and to which molten solder will not stick*

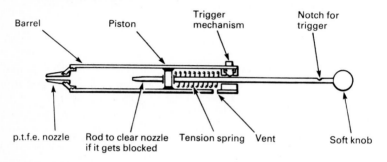

The principle is very simple, rather like a back-to-front bicycle pump. The plunger is pushed down against the spring, where it is locked by the trigger. The soldered joint to be released is then heated with a soldering iron, the nozzle of the sucker applied to it, and the trigger pressed. The spring forces the plunger up the tube, sucking the molten solder up after it and also cooling it instantly. Some solder suckers have a double-sprung plunger to prevent it flying out and hitting you in the face when you press the trigger.

Hand tools

A range of small hand tools is used for electronics work, but there are few specialised tools. A variety of long- and short-nosed pliers should be available, as well as small and medium side-cutters. Several screwdrivers, both flat and cross-point, are essential. The only specialised tools that are often found are *wire strippers*, for removing the insulating plastic from connecting wires; *trimtools*, small plastic screwdrivers for adjusting the cores of inductors; and perhaps a miniature electric drill and magnifying glass.

In some high-quality assembly and prototype work, an assembly technique known as *wire wrapping* is used; there are a number of special devices to be used, for wrapping and unwrapping wires. Briefly, wire wrapping consists of making a connection by wrapping a wire tightly round a hard square post with sharp corners. This results in a joint with good mechanical and electrical properties, made without heat.

4.2 TEST EQUIPMENT

The multimeter

If the electronics engineer is to have any idea at all of what is going on in circuits, a range of instruments must be available for measuring the various quantities involved. The *multimeter* is the electronic engineer's most basic tool, and has been developed as a quick and easily used means of measuring basic electrical quantities.

A multimeter consists of a measuring device, plus associated circuits to select a suitable range and function.

The range and function controls are made to be as convenient as possible for ease of use. A good multimeter might have the following functions and ranges:

Direct voltage	*Alternating voltage*
0–0.1 V	
0–1 V	
0–10 V	
0–100 V	ranges as for d.v.
0–1000 V	
0–5000 V	

Direct current	*Alternating current*
0–100 μA	
0–1 mA	
0–10 mA	
0–100 mA	ranges as for a.c.
0–500 mA	
0–5 A	

Resistance

0–100 Ω
0–1 kΩ
0–100 kΩ
0–10 MΩ

Multimeters from different manufacturers will have rather different ranges, but the above gives an idea of what is average in a good-quality meter.

There are other factors, besides ranges, which need to be taken into account. One is the *input resistance*. It is no good having a meter which reads voltage if the meter itself imposes a heavy load on the circuit that it is measuring and changes that voltage. Similarly, a meter to measure current should place as little series resistance as possible in the circuit being measured, or the measurement will prove unacceptably lower than the current that flows without the meter. When measuring resistance, which requires a current to be passed through the thing being measured, the current should not be too great or the applied voltage too high.

All the above factors are a function of price—in general, the more you pay, the better the specification.

Analogue and digital

Traditionally, multimeters were made using high-quality jewelled meter movements, made to the finest engineering tolerances. Today, such meters are still available but there is an option to use a digital display instead.

For most purposes, the digital meter is the better option. A typical digital meter has a '$3\frac{1}{2}$ digit' display, meaning that it can display any number between 0000 and 1.999. This type of display is used because it virtually doubles the range with only a little extra electronics.

The digital meter is usually more accurate and more robust than the mechanical 'analogue' meter, price for price. It is, however, less suitable for measuring continuously changing quantities than the analogue meter— where the meter needle can swing to and fro with the changes to indicate roughly what is happening (following a 1 Hz waveform, for example), the digital meter becomes a confusion of changing figures.

For details of the principles of the analogue meter, see Chapter 17; and for details of digital meters, see Chapter 23.

The oscilloscope

The cathode ray oscilloscope (CRO) is probably one of the most useful items of test equipment, after the multimeter. The CRO uses a cathode ray tube (see Chapter 5) to display the waveform of an electrical voltage. The tube is a precision device, and is designed so that the electron beam (producing the spot on the screen) can be deflected by an accurately controlled amount. The screen is ruled with a grid of squares, called a *graticule*, and is arranged like a graph, with a vertical y-axis and a horizontal x-axis. A typical CRO screen layout is given in Figure 4.6.

The grid is conveniently ten squares in both directions. The (horizontal) x-axis is calibrated in *time*, and the y-axis is calibrated in *voltage*. Both the time scale and the voltage scale are adjustable, by means of range-setting controls. A small general-purpose oscilloscope is illustrated in the photograph in Figure 4.7.

fig 4.6 *an oscilloscope graticule*

Centre lines marked in tenths   10 x 10 grid

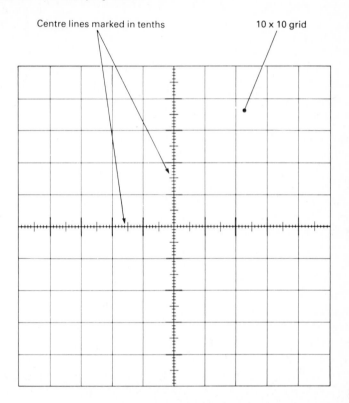

fig 4.7 *a typical small oscilloscope*

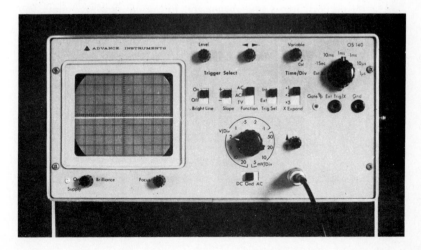

The x-axis is calibrated in 'time per division', making the spot scan the screen at a rate controllable from a maximum speed of 1 μs per division to a minimum of 0.1 s per division. The range-setting control is in the top right of the front panel.

In the centre of the lower part of the panel is the control for setting the range of the y-axis. This is variable, in this particular model, in twelve steps from a maximum sensitivity of 5 mV per division to a minimum sensitivity of 20 V per division. The control for this is in the middle of the lower part of the photograph.

A graph of almost any electrical waveforms can be displayed on the screen, providing an invaluable aid to understanding circuit function or to servicing most electronic equipment.

A most important feature to be found on the oscilloscope is the *trigger* or *synchronisation* system. Circuits in the instrument detect the level of the signal being measured, and trigger the sweep of the x-axis (at the set rate) at a predetermined point in the waveform. Because the sweep is always triggered at the same point on the waveform, the picture on the screen is automatically 'locked' in place on the screen, and does not drift from side to side with small changes in frequency. Provision may be made to trigger the sweep of the x-axis from an external source, perhaps derived from the circuit under measurement. Additionally, there may be special-purpose filter circuits built into the CRO to select, for example, television synchronisation signals (see Chapter 15) and lock the picture on to them. A range of such facilities can be seen in the photograph.

The CRO can, of course, be used as a measuring instrument, as the 'graph' on the screen is accurately calibrated. Voltage can be measured by reference to the y-axis, and frequency or pulse-widths by reference to the x-axis.

A good oscilloscope is not cheap, and the more expensive units will have better accuracy, more facilities and greater range. Oscilloscopes designed to work at high frequencies also tend to be expensive, as do CROs designed specially for digital work. A good one may cost as much as a family car.

The more expensive models have a *double-beam* tube, which displays two waveforms simultaneously. The x-axes are synchronised with each other, but the y-axes are controlled by independent inputs. This facility permits of comparison of waveforms—to test a divider or pulse-shaping circuit, for example.

The signal generator
The signal generator is an instrument for producing a waveform at a known frequency, and having a known shape. Most signal generators can provide a *sinewave* output or a *squarewave* output, with a peak-to-peak voltage

ranging from a few millivolts to a few volts. Frequencies from a few Hz to a few MHz are standard in an inexpensive instrument, and it is often possible to produce an output that is a radio frequency carrier, modulated with an audio signal (see Chapter 15).

The better signal generators will also produce output pulses that are compatible with digital logic circuits (see Chapters 20 and 21), with both the pulse-width and frequency independently variable.

The transistor tester

Oddly, it is seldom necessary to test a transistor. Reference to the manufacturer's specifications or a simple substitution are usually all that is necessary. It is, however, possible to buy instruments that will measure a transistor's parameters accurately. Such instruments are, in fact, seldom used, except in the laboratory. Somewhat more useful are *in-circuit transistor testers*, which can perform a 'go/no go' test on a transistor without the necessity for removing it from the circuit it is fitted into. Such devices are useful servicing aids.

4.3 SAFETY

Safety in using test equipment

This refers to the safety of the equipment, not the safety of the wielder of it! It is always possible to devise a way of damaging or destroying completely an item of test equipment, despite all the safety devices the manufacturers have built into it. Often such ways are discovered accidentally by engineers and students who should know better.

Always, always, make sure that you have selected a suitable range before connecting a meter or CRO to a 'live' circuit. Don't expect to be able to set a meter to read full scale 10 mA and then connect it across a mains supply without a flash and a bang and an expensive repair bill. Oscilloscope input amplifiers can be wrecked in the same way, by selecting a range that is far too low—and the repair bill for even a cheap oscilloscope will amaze you—adjusting and recalibrating a CRO is labour-intensive and therefore costly.

Remember that while little damage will be done by connecting a meter set to read volts in series with a circuit (expecting to read current), you must be very careful not to select a current range and then try to read a voltage. The low resistance of the meter will result in all the available current flowing through the meter . . . if the meter has a fuse or automatic cutout, you *might* be lucky . . . . The same applies to measuring resistance. Never try to use a resistance range on a circuit that is live. At best the results will be suspect, at worst you will destroy the meter. Remember, too, that 'live' can mean a charged electrolytic capacitor in a circuit that is disconnected from its power supply.

Finally, it is wise to remember the remark of the salesman who, when asked the working life of his company's latest digital multimeter, replied 'Until you knock it off the workbench.'

General safety

The trouble with electricity is that you can't see it, smell it, or hear it. You can, however, feel it. And in the right (or wrong) conditions, it can kill you.

In the UK the Health and Safety at Work Act 1974 places the responsibility for people at work on the employer *and* the employee. In other words, although your employer has a duty to make sure that working conditions are as safe as is reasonably possible, it is up to you to take care of yourself.

Most serious injuries and deaths from electric shock occur from contact with the mains electricity supply, which in the UK is around 240 V a.c. The *live* wire of the mains supply is at a high voltage with respect to earth, and if you are touching the ground you can receive a fatal electric shock by touching the live wire *only*. Because electric shock kills by temporarily paralysing the heart muscles, the greatest danger occurs when the current flow is across the chest—either from one arm to the other, or from either arm to either leg.

The resistance of a human body depends mostly on the resistance of the skin. If the skin is damp, either from water or from perspiration, the resistance is dramatically reduced. In such circumstances the chances of any electric shock being a fatal one are greatly increased, so you should never operate or touch electrical equipment with wet hands.

The secondary effects of electric shock can also be dangerous. A television engineer was seriously hurt when he was fitting a loft aerial. His assistant plugged the aerial into a faulty television receiver (it had been modified by an amateur 'expert') and the aerial became live to the mains. The shock threw the engineer off the joist he was standing on; he fell through the plasterboard ceiling and came down astride an open door.

Electronics engineers run special risks when dealing with apparatus that uses high voltages. Colour-television receivers use very high voltages— up to 30 kV—at relatively high currents. Contact with a colour television EHT (extra high tension) connection is usually fatal.

Apparatus designed to be connected to the mains should either be *earthed* or double insulated. These precautions greatly improve the safety of the equipment, but *not* if you take the covers off for working on the insides. Figure 4.8 compares the way the two methods of protection work for faulty equipment.

fig 4.8

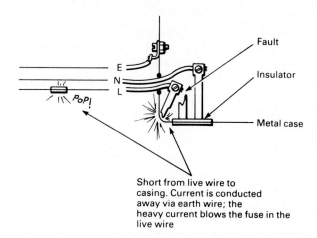

Fault

Insulator

Metal case

Short from live wire to casing. Current is conducted away via earth wire; the heavy current blows the fuse in the live wire

(a)   *an earthed appliance*

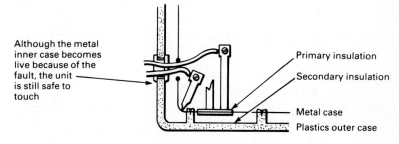

Although the metal inner case becomes live because of the fault, the unit is still safe to touch

Primary insulation

Secondary insulation

Metal case

Plastics outer case

(b)   *a double-insulated appliance*

An *isolation transformer* provides a measure of safety when working on equipment that requires mains voltage. The isolation transformer is a transformer with two separate windings, insulated from each other to a high standard. The windings have a turns ratio of 1 : 1, so that the output voltage is the same as the input voltage. Safety is improved because neither of the two supply wires is live with respect to earth. Touching just one or the other will not result in a shock. The voltage is still just as high, however, so all the usual precautions should be taken.

Take extra care when switching off or isolating circuits, particularly mains. Anyone can make mistakes. I am lucky to have written this book, for a few years ago I was working on a mains line. The line was controlled by a double-pole isolating switch of approved design. Before starting, I checked that the switch was off, double-checked in fact. I then took my cutters and carefully cut the wire *on the wrong side of the switch*. It cost me a good pair of cutters, but it could easily have cost a lot more.

Fire is always a danger. Electrical equipment of all types dissipates heat, and under fault conditions the heat output of a component can rise high enough to cause a fire. I once saw a colour television with most of its insides and part of the cabinet completely devastated by a fire caused by a faulty capacitor. The capacitor short-circuited and overloaded a $1 \text{ k}\Omega$ $\frac{1}{2}$ watt resistor. The resistor overheated and melted a plastic insulating sleeve over high voltage connections to the back of the tube. This caused a massive short-circuit, which set fire to the whole receiver. The owner was fortunate that the fire confined itself to the television.

Various types of fire extinguishers are available—and it should go without saying that water should never be used in an attempt to put out an electrical fire. Fires are extinguished by removing the combustible material—seldom possible—or by removing the oxygen supply, or by cooling the burning material below the point at which combusion can be sustained.

Foam extinguishers work by smothering the fire in a foam containing the inert gas carbon dioxide and depriving it of oxygen. However, they contain *water* and are unsuitable for electrical fires.

Carbon dioxide extinguishers contain liquid carbon dioxide gas, which smothers the fire with inert gas and starves if of oxygen. This type of extinguisher is good for fires of all types, especially electrical fires. Although the gas is very cold as it comes out of the nozzle, it has little cooling effect by the time it reaches the flames and the fire may re-ignite if the gas is blown away or begins to dissipate. Carbon dioxide extinguishers are also suitable for burning liquids.

Dry powder extinguishers are useful for electrical fires, especially fires that are fairly confined—such as a burning item of equipment. The dry powder does no damage and can be brushed off, so there is more chance of salvaging apparatus that has been extinguished with this type of appliance. Dry powder extinguishers are not suitable for large fires.

Fire blankets are very effective. They are usually made of aluminised glass-fibre cloth and are simply placed over the burning object to deprive the fire of oxygen. This does mean going right up to the fire, so be careful.

Sand also works quite well, but should not be used on burning liquids as it simply spreads them around.

It is important that you only tackle a fire (i) if you think it is small enough to put out—don't overestimate the power of even large extinguishers; (ii) if there is no personal danger; (iii) after you have raised the alarm. If a piece of equipment, or even a whole room is ablaze, *shut the door*. This will slow the fire's spread, even if the door is an ordinary plywood interior door. It will also slow the spread of smoke and possible toxic fumes, potentially more dangerous than the fire itself.

As a preventative measure, don't leave equipment running if there is nobody about, and make sure that all fire doors, fire escape routes and fire appliances are in usable condition.

4.4 FIRST AID

This is not a book on first aid, and there are many excellent books on the subject available at bookshops and chain stores. Buy one and read it.

The special hazard in electronics and electrical work is shock. Remember that shock kills by paralysing the heart and sometimes the respiratory muscles. If you can keep a victim alive for a few minutes, the chances of complete recovery are good.

When attempting to help someone who is the victim of electric shock, *make certain that the cause of the shock has been removed before you touch the victim*. Otherwise you may become the second victim.

If there is a mains switch nearby, cut the power. If this is not possible, use an *insulator* to get the person away from the power. A broom handle, wooden or plastic chair, improvised rope (your shirt, a rug, jacket, etc.), will do. If the victim is *holding* the source of the shock, it may take force to make him let go, since the electricity contracts the muscles of the hand.

If the victim is unconscious, check to see if breathing continues. If not, mouth-to-mouth artificial respiration should be begun. If the victim does not start to breath spontaneously, check for heartbeat—listening to the bared chest is the easiest way—and if you cannot detect a heartbeat or see a pulse beating in the victim's neck, then begin cardiac massage. The techniques are described in any good first-aid book. Remember that the brain cannot survive more than three or four minutes without a supply of oxygenated blood, so don't waste time.

If all this makes electronics sound very hazardous, the good news is that modern circuits tend to use low voltages rather than high voltages. With a few exceptions—portable televisions and electronic flashguns, for example—battery-powered equipment is unlikely to give the careless engineer a shock. But it pays to be careful. If in doubt, don't touch.

QUESTIONS

1. You want to measure, approximately, the frequency of a signal. A frequency measuring meter is not available. What item of test equipment would you use?

2. Unfortunately the item of test equipment used in the above question suddenly bursts into flames. Describe the correct procedure for dealing with this emergency.

3. Why is double-insulated equipment safe without an earth connection?

4. Which of the following metals might prove difficult to solder: tin, copper, aluminium, brass, gold, steel.

5. What are the maximum and minimum voltages that can be read on a good $3\frac{1}{2}$ digit digital voltmeter?

PART II
LINEAR ELECTRONICS

THERMIONIC DEVICES

Thomas Alva Edison is chiefly known as the inventor of the phonograph, and as one of the inventors of the electric lamp. It is less well known that Edison *nearly* invented the *thermionic diode*—the 'breakthrough' device that heralded the beginning of the science of electronics.

In the early 1880s Edison was trying to improve his electric lamp. One of the problems he had was in preventing the inside of the glass envelope going black and obscuring the light. He realised that the filament of the lamp was evaporating, and wondered if he could use a wire grid to intercept the material before it got to the glass. The modified lamp was based on the design in Figure 5.1.

Unfortunately it didn't work, but during his experiments Edison did try the effect of applying an electric voltage to the grid. He observed that if the grid were made negative with respect to the filament, nothing much happened, but if he made it positive with respect to the filament, he could draw a substantial current from it. This was mildly interesting but was not going to help him with the electric lamp, so he called it the 'Edison effect' and filed the idea for future development . . .

fig **5.1** *Edison's modified lamp*

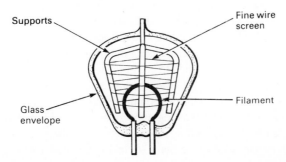

It was left to John Ambrose Fleming, in 1904, to do the essential pioneering work in electronics, and the device which carries his name, the *Fleming diode*, bears tribute to his work.

5.1 THE THERMIONIC DIODE

The simplest form of Fleming diode is almost identical to Edison's modified lamp, and is illustrated in Figure 5.2.

fig **5.2** *a Fleming diode*

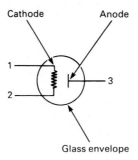

The cathode is just like a lamp filament, though it is treated with thorium to increase the number of electrons that can be driven off when it is heated. Facing the cathode is the *anode*, a metal plate connected to an outside terminal. The two parts are sealed inside a glass envelope in which there is a vacuum. It is easy to see how the Fleming diode, or *thermionic diode*, works.

When the cathode filament is heated by passing a current through terminals (1) and (2) in Figure 5.2 electrons are driven off the filament. These free electrons are separated from their parent atoms by the heat. If the anode is made negative compared with the cathode, the electrons are repelled by the anode (*like repels like*; remember that the electrons have negative charges) and no current can flow through the diode.

On the other hand, a positive charge on the anode will attract electrons given off by the cathode, and they will move through the vacuum inside the envelope; this movement of electrons constitutes an electric current. Figure 5.3 shows a demonstration circuit for the thermionic diode.

The single cell provides heating current for the cathode, but otherwise plays no part in the operation of the circuit. Consider the effect of an alternating voltage applied across *A–B* in Figure 5.3. While *A* is positive with respect to B a current will flow through the load resistor, R_L, but if B is more positive than A, no current will flow. Figure 5.4 compares the applied voltage with current flowing through R_L.

This is the principle of *rectification*—turning a.c. into d.c. The original Fleming diode was developed into the modern rectifier valve, shown in Figure 5.5.

The major change is the fact that the cathode is indirectly heated. This separates the heating and electron-emitting functions. The heater filament

fig **5.3** *demonstration circuit for a thermionic diode*

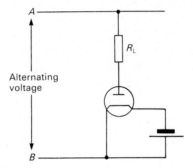

fig **5.4** *rectification of an alternating voltage supply*

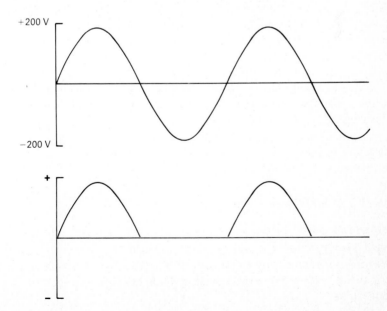

fig 5.5 *construction of a modern thermionic diode valve*

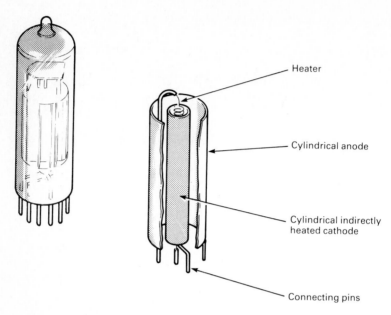

Heater

Cylindrical anode

Cylindrical indirectly
heated cathode

Connecting pins

is inside a narrow tube of thin metal, but nowhere does it come into contact with the tube. The filament gets red hot, and this heats the tube (the cathode) which emits electrons. The main advantage is that the heater is electrically isolated from the cathode, and in a system involving several diodes (and other thermionic devices) the heaters can all be run from the same power supply, which can be a.c. for cheapness. The temperature of the heaters will fluctuate with the changing current of the supply, but that of the cathodes will remain relatively constant.

Another obvious (but important) development was the idea of putting all the connections at one end of the valve, so that it could be plugged into a base for easy replacement when the heater burned out.

Thermionic diodes like the one in Figure 5.5 were in regular use in all types of equipment until the end of the 1960s, when they began to be superseded by *solid-state* diodes.

5.2 THE THERMIONIC TRIODE

The *triode valve* was invented by Lee De Forest in 1910. It is constructed like a Fleming diode but with an extra electrode, consisting of a grid of fire wires, interposed between the cathode and the anode. The basic layout is shown in Figure 5.6, which, incidentally, is the circuit symbol for the triode valve.

fig 5.6 *circuit symbol for a thermionic triode*

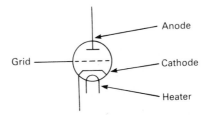

With the control grid unconnected, the triode behaves like a diode; but if the grid is held at a small negative potential relative to the cathode, the current flowing through the valve is reduced. It is easy to see why this is so; electrons travelling from the cathode to the anode are repelled by the charge on the grid and prevented from passing. Make the grid slightly more negative still, and the current that can flow from the anode ceases altogether.

For all practical purposes the grid takes no current at all, and the maximum voltage required on the grid to stop all current flow from the anode is substantially less than the anode voltage. Between certain limits (which depend on the valve construction and on the circuit in which it is used) the voltage at the anode of the valve is proportional to the grid voltage, but larger. Thus the valve can be used to amplify a signal applied to the grid, by increasing the amplitude (size) of the signal without changing its form.

Figure 5.7 shows a triode valve in a simple amplifier circuit. The two resistors R_g and R_c are involved in keeping the grid at a more negative potential than the cathode. The output waveform as shown in the figure is the same *shape* as the input waveform, but has a larger voltage change— it is amplified. A small voltage applied to the grid controls a much larger voltage. *The principle of a small voltage or current controlling a larger voltage or current is one of the most fundamental concepts in electronics.*

There are many types of valve apart from the triode. Almost all the designs are intended to improve the basic efficiency or to compensate for some undesirable characteristic. And almost all these improvements made the valve more complicated, and so more expensive. Valves are still used today—but not much. In all but a few specialist applications, the old 'vacuum-state' electronic devices have given way to solid-state equivalents.

There is only one really common thermionic device in use today. We will look at this next.

fig 5.7 *a triode valve in a simple amplifier circuit (the heater connections are not shown)*

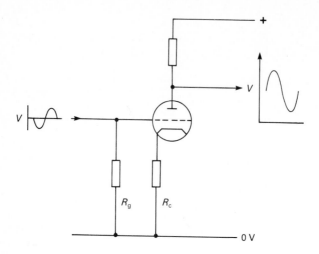

5.3 THE CATHODE RAY TUBE

Most of us are familiar with the appearance of the cathode ray tube (CRT). More people spend more time looking at CRTs (the front, at least) than at anything else.

In the second half of this chapter, we can examine the inside of the TV tube—black-and-white version! Figure 5.8 shows the basic design of the CRT. Compare it with the triode in Figure 5.6, and you will see that it has more than a passing resemblance to a very large valve.

The cathode is indirectly heated, and is designed to emit electrons from a fairly small source. In front of the cathode is a control grid, in the form of a cylinder. In front of the control grid are *two* cylindrical anodes, carefully designed to act as a 'lens' when the proper voltages are applied. The object of these anodes (termed *first* and *second* anodes) is to concentrate the electrons into a beam, aimed at the front of the tube. Figure 5.9 shows the way that the electric field produced by the two cylindrical anodes acts as an *electron lens* to concentrate the beam of electrons, just like a glass lens can be used to concentrate a beam of light.

Next there is a third anode (usually called the *focus anode*); voltage applied to this can be finely adjusted to bring the electron beam to a sharp focus on the screen. Lastly, there is a *final anode*, which is connected to a graphite coating applied to the inside of the flared part of the tube. All the

fig **5.8** *cross-section through a cathode ray tube*

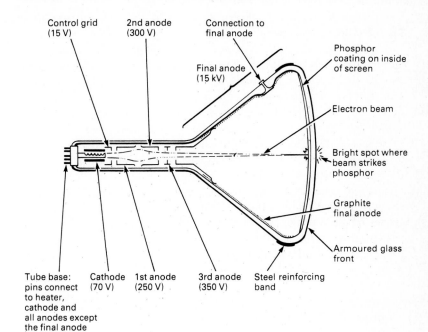

fig **5.9** *the electric field between the first and second anodes concentrates the beam of electrons, in much the same way that a lens concentrates light*

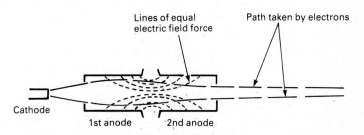

flared part of the tube is thus at the same potential as that of the final anode—in fact the whole of the front part of the tube could be said to *be* the final anode, as it is all connected together.

The front part of the tube—the screen—is coated on the inside with a material called *phosphor*. This has the property of glowing brightly when struck by electrons (phosphors are also used on the inside of fluorescent lamps to make them glow). Phosphors can be made almost any colour—a television tube would use one that glows white, while computer terminals often feature a green screen.

Operation of the tube is straightforward in principle. Electrons are emitted by the heated cathode, formed into a beam by the first two anodes, and focused by the third anode. The beam strikes the phosphor on the screen to produce a bright dot. The *beam current*, and thus the brightness of the dot, can be controlled by the voltage applied to the control grid.

A television tube is relatively long, so the electrons need a lot of energy to persuade them to travel down the tube. Moreover, they must have enough energy left to make the phosphor glow brightly when they reach the screen. For these reasons, the voltages associated with CRTs are high. Figure 5.8 indicates typical voltages; the voltage on the final anode is always very high. It is so high, in fact, that the final anode connections cannot be made to the pin connectors at the back of the tube; this would pose insurmountable insulation problems. Instead, the high voltage connection is made to a special plug on the side of the flared part of the tube. A thick, heavily insulated wire to the final anode is a prominent feature of any TV receiver.

5.4 BEAM DEFLECTION

So far we have described a CRT that produces a single dot in the centre of the screen. The brightness can be controlled by the grid, but we have to use some extra parts to move the dot around the screen. The most common method is to use *magnetic deflection*. Electrons are easily influenced by magnetic fields, so a suitably designed system of coils, slipped over the outside of the tube neck, can be used to bend the electron beam, causing the dot of light to move up and down and from side to side. Two separate sets of coils are used, for vertical and for horizontal deflection. Television receivers and computer terminals always use magnetic deflection, but the *oscilloscope*, for example, uses another form of beam deflection, *electrostatic deflection*.

The oscilloscope is a measuring instrument that displays electrical waveforms on a CRT. The screen is marked with a grid and time and voltage can be read off the screen. An accurate and linear method of beam deflection is needed—electrostatic deflection is the answer. The tube is fitted internally with parallel horizontal and vertical plates, mounted in front of the focus anode. A high voltage applied to opposite plates will deflect the electron beam, and the degree of deflection on the screen is proportional to the applied voltage: ideal for a measuring instrument. Figure 5.10 shows the relative proportions of a television picture tube with magnetic deflection and an oscilloscope CRT.

fig 5.10 *comparison of designs of cathode ray tubes—the upper one shows a television tube, and the lower one an oscilloscope tube*

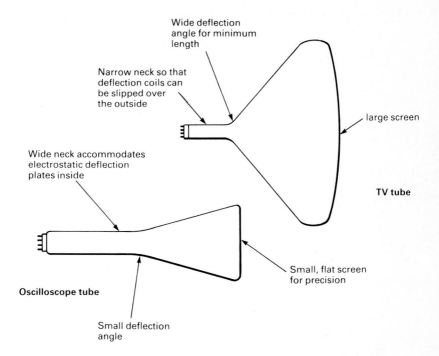

Wide deflection angle for minimum length

Narrow neck so that deflection coils can be slipped over the outside

large screen

Wide neck accommodates electrostatic deflection plates inside

TV tube

Oscilloscope tube

Small, flat screen for precision

Small deflection angle

The wider deflection angle of the TV tube is designed to make the tube shorter for a given screen size. Design of the deflection coils and the circuits driving them is more difficult for wide-angle tubes, but the convenience (and better saleability) of shallow television receivers makes it worth the effort. The oscilloscope, on the other hand, is a piece of test equipment— accuracy is the important factor, and the length of the instrument is unimportant.

Colour television tubes work on the same principle, but are considerably complicated by the need to produce three independently controllable, perfectly superimposed pictures in red, green and blue! Colour television picture tubes are given further explanation, along with TV systems, in Chapter 15.

QUESTIONS

1. Why does the cathode of a Fleming diode have to be heated?

2. In a television tube, electrons are fired at the front of the tube by an electron gun. What makes the front of the tube light up when the electrons strike it?

3. State the two different types of deflection system used in a CRT.

4. What voltage would you find on the final anode of (i) a monochrome, (ii) a colour TV tube? Is the voltage high enough to be dangerous?

5. What controls the beam current in a CRT? What effect does changing the beam current have on the screen?

SEMICONDUCTORS

In the first chapters of this book we have used a model of the atom origin-ally designed by Niels Bohr. It is easy to picture the Bohr atom, with its hard, bullet-like electrons hurtling round the massive nucleus just like planets orbiting a sun in a tiny solar system. The work done by Werner Heisenberg in the late 1920s showed that this model is unfortunately further from reality than we might find comfortable and that the atom is actually a rather fuzzy and uncertain thing, not at all like Bohr's micro-miniature solar system. This is not a book about atomic physics, so it is unnecessary to look too closely at the construction of atoms—but it is important to realise that there are 'rules' that appear to govern the behaviour of atoms and their component parts. Many of these rules seem contrary to what we would expect; but our intuition is necessarily based on our experience of the behaviour of objects much larger than atoms and electrons.

Consider an individual atom of an element—silicon is a useful example. The atom consists of a central nucleus surrounded by a cloud of electrons, which can be represented diagrammatically in Figure 6.1.

The electrons arrange themselves into three orbits, or 'shells'; although the diagram shows the electrons in a flat plane, the orbits actually occupy

fig 6.1 *one model of the silicon atom*

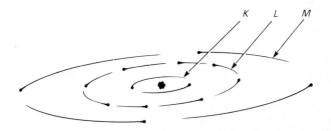

a spherical 'shell'. The shells are given letters, starting with K for the inner-most shell, then L and M, and if there are more than three shells N, and so on. Each shell can hold a specific number of electrons, two in the K shell, eight in the L shell, and eighteen in the M shell. As we build up models of different atoms, the shells are filled from the orbit nearest the nucleus, so silicon (which has fourteen electrons) has full K and L shells (two and eight) and the remaining four electrons in the M shell.

This seems straightforward enough, but if we look closer at the atom we find a little of Heisenberg's 'fuzziness' beginning to creep in. One of the rules governing the behaviour of electrons in a system (a system means an atoms or group of atoms) states that no two electrons can be at precisely the same energy level.

What does 'energy level' mean in this context? Go back to the solar system analogy and imagine a spacecraft orbiting in the K shell. Run the engines to increase its speed, and it will move out to a more distant orbit, perhaps even as far as the L shell. Subtract energy from the spacecraft by allowing some energy to dissipate as heat (it's a low orbit, in the outermost fringes of the atmosphere, and is subject to a little drag!) and it will drop into a lower orbit. Thus it is clear that the greater the energy possessed by the spacecraft, the higher—further from the nucleus—will be its orbit.

If no two electrons can have the same energy, it follows that no two electrons can orbit at exactly the same distance from the nucleus. It also follows that the shells consist of more than one possible orbit. The L shell, with its eight electrons, must consist of at least eight different orbits, all close to one another but not the same. Can we say how these orbits are arranged, and which of the possible orbits are in fact occupied by an electron? Unfortunately we cannot. Another of the rules governing the behaviour of electrons says that we cannot know the speed, position and direction of an electron all at once. We can never say for certain the whereabouts of the eight electrons that form the L shell at any particular instant; all we can do is say where there is the greatest *probability* of their being located. Compare part of the 'Bohr' orbit in Figure 6.2a with the 'Heisenberg' version in Figure 6.2b.

The degree of shading in Figure 6.2b represents the degree of probability of an electron being in that particular orbit. This is all we can say about

fig **6.2** *two models showing sections of an electron's orbit*

(a)   (b)

the electrons, not because of any limitations in our measuring equipment, but because of the very nature of electrons. This rather odd fact about electrons is one of the more important discoveries to come out of modern quantum physics.

But Bohr's model is reliable in that there are specific regions (or shells) that the electron can occupy. It is not possible for electrons to orbit *between* the shells. We can redraw Figure 6.1 to show the probabilities of electrons being in any particular orbits (see Figure 6.3).

fig **6.3** *atomic model showing energy bands and forbidden gaps*

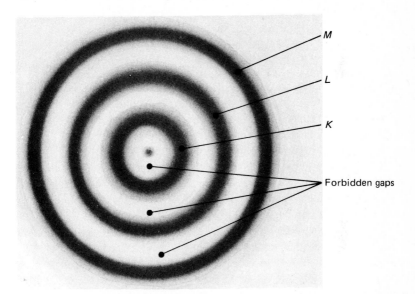

It now seems more realistic to call the shells 'energy bands', as they represent a range of possible electron energies. The higher the energy, the further it will be from the nucleus. Electrons may not orbit in the *forbidden gaps* between the shells. The outer band is called the *valence band* and is the only band that may not be completely full.

If we take a section through this model of the atom, from the nucleus to the outermost band, we can represent this by a diagram showing just a small part of each orbit—like the one in Figure 6.4. We can use this as an energy-level diagram for our atom, since the bands represent electrons having increasing energy as we go up the diagram. The degree of 'fuzziness' of the bands depends on the temperature, and it is normal to show energy-level diagrams for a temperature at (or very near) absolute zero.

It is possible to make an imaginative jump and use one energy diagram

fig 6.4 *a section of the orbits shown in Figure 6.3*

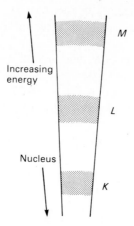

to represent the average of a very large number of atoms in a system. This is what we do to describe the operation of semiconductor devices. Such energy-level diagrams are, as we shall see, very useful aids to understanding quite subtle atomic interactions. But let us refine the diagram a little more before using it. . . .

It is useful to know, for example, that the inner shells of the atom are not generally involved in any of the interactions that occur in electronics. This simplifies the energy-level diagram, for it means we can simply leave out the lower bands. Figure 6.5 shows an energy-level diagram for a piece of silicon—just showing the valence band. There is a vast number of possible orbits—all different—making up the band. Many or most of these possible orbits will be unoccupied. It is even possible to imagine a band which is completely unoccupied! Such a band is still there, theoretically at least, as it defines the probable positions of any electrons that somehow gain so much energy that they leave the valence band. So it is useful to add another band to the energy-level diagram, an 'empty' band beyond the valence band. This band is called the *conduction band*, and both it and the valence band are shown in Figure 6.6.

fig 6.5 *possible orbits in the valence band*

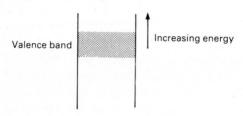

fig 6.6 *energy-level diagram showing the valence and conduction bands (the conduction band need not have any electrons in it)*

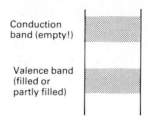

Conduction band (empty!)

Valence band (filled or partly filled)

6.1 CONDUCTORS AND INSULATORS

We can use energy-level diagrams to explain why conductors conduct and insulators do not. Compare the energy-level diagrams for copper and sulphur (Figure 6.7).

In Figure 6.7a there is no forbidden gap between the valence and conduction bands. Electrons can move freely from band to band and there is no barrier to electron movement—a small increase in energy can move an electron into the conduction band. Once in the conduction band, the electrons are not bound to the structure of the atoms, and are free to drift through the material as an electric current.

Sulphur is different. The sulphur atom has a rather large forbidden gap between the valence and conduction bands. A moderate increase in energy will not be sufficient to move an electron out into the conduction band, but would only lift the electron as far as the forbidden gap. Since electrons cannot exist in the forbidden gap, they will not be able to accept such an energy increment, and will stay in the valence band. The conduction band will remain empty, and no current can flow.

fig 6.7 *energy-level diagrams of a typical conductor (copper) and a typical insulator (sulphur)*

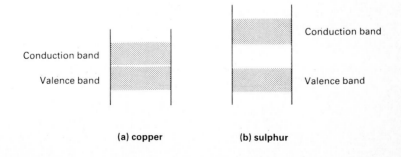

Conduction band

Valence band

Conduction band

Valence band

(a) copper   (b) sulphur

6.2 INTRINSIC SEMICONDUCTORS

Figure 6.8 shows an energy-level diagram for germanium. The forbidden gap between the valence and conduction bands is very small and it requires only a little added energy, such as the thermal energy available at normal room temperature, to cause a few electrons to jump into the conduction band. Germanium will therefore conduct electricity, but poorly—less than a thousandth as well as copper, in fact. Germanium is called an *intrinsic semiconductor*, since it naturally has properties between those of conductors and non-conductors. The conductivity of germanium is strongly affected by temperature, as you might expect. The higher the temperature, the more the energy bands blur and expand, and the smaller the forbidden gap becomes.

fig **6.8** *energy-level diagram showing a typical intrinsic semiconductor*

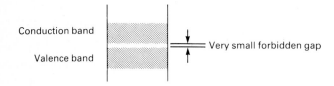

6.3 CHARGE CARRIERS IN SEMICONDUCTORS

We have seen how electrons can drift through a material in the conduction band and how this forms a flow of electric current. Electrons that are free to move about in this way are called *charge carriers*, because they carry electric charge through the material. But there is another mechanism by which charge can be carried through a substance, and we can see how this operates by looking at an energy-level diagram showing what happens when an electron makes the transition from the valence band to the conduction band (see Figure 6.9).

fig **6.9** *when the electron moves from the valence and into the conduction band it leaves a hole in the valence band*

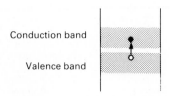

The movement of an electron from the valence to the conduction band provides a 'free' electron in the conduction band, but it also leaves a *hole* in the valence band, i.e. a gap into which an electron might easily move. The hole could be filled by another electron in the valence band, but this would leave a hole somewhere else, which could be filled by another electron, leaving another hole in the valence band, and so on, and so on. The creation of the electron-hole pair that results from an electron moving from the valence to the conduction band actually allows electrons to move about freely in the valence band as well as in the conduction band, though it is conventional to think of movement of electrons in the conduction band and of movement of holes in the valence band. Figure 6.10 shows clearly how a movement of holes is really the same as a movement of electrons in the opposite direction—both ways of looking at it are equally valid. Just as the electron carries one unit of negative charge, so the hole can be said to carry one unit of positive charge, which will in the right circumstances exactly cancel out the negative charge on the electron.

It is reasonable to ask at this stage just why we should bother to describe such a relatively complicated mechanism when the net result is simply an electron drift through the conduction and valence bands. The answer is that in the artificial semiconductors used in electronic devices we can

fig **6.10** *the movement of electrons in one direction is the same as move-ment of holes in the other direction; this can be demonstrated using a chess board with a row of 7 pawns—(a) shows the initial arrangement: one pawn is moving to the right in each stage of this diagram, (b) to (f). Although no electron has moved more than one place, the hole has drifted from the extreme right of the row to the extreme left*

deliberately introduce an excess of either holes or electrons into a substance, and it is often one or the other mechanism which dominates the conduction inside the material. Artificial semiconductors are known as *extrinsic semiconductors*.

6.4 EXTRINSIC SEMICONDUCTORS

We start with a cheap, plentiful semiconductor like silicon, purify it until it is absolutely pure, and then add a tiny amount of another substance— one part in 100 000 000 of phosphorus, for example. The resulting mixture is then made into a single perfect crystal. Because of its atomic structure, the phosphorus atom fits nicely into the crystalline matrix of the silicon, but with one electron left over in its outer shell, for it has five and not four electrons in its valence band. This extra electron is 'spare' to the structure of the crystal, and moves easily into the conduction band. An energy-level diagram for a crystal of silicon 'doped' with phosphorus is shown in Figure 6.11. We show the extra electrons in the conduction band as minus signs. Such a doped semiconductor is called an *n-type* semiconductor, the *n* representing 'negative' to indicate that the material has extra (negatively charged) electrons.

fig **6.11** *energy-level diagram for an n-type semiconductor*

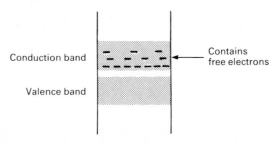

Alternatively we could add to the silicon crystal matrix a few parts per million of boron. The boron atom has only three electrons in its valence band, and it too drops neatly into the silicon crystal structure. But it leaves a hole. This hole is easily filled by an electron from the valence band, leaving a corresponding hole in the valence band—and we have already seen that such holes act as positive charge carriers. Silicon doped with boron is therefore known as a *p-type* semiconductor, indicating that it has extra (positive) holes in its structure that can work as charge carriers. A suitable energy-level diagram is given in Figure 6.12. Most of the holes appear at the top of the valence band, with their concentration falling off nearer to the nucleus. In Figure 6.11 the electrons had their highest concentration at the lower levels.

fig 6.12 *energy-level diagram for a p-type semiconductor*

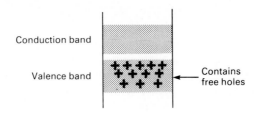

This seems less strange when you realise that a hole needs *more* energy, not less, to get nearer the atomic nucleus. The concentration varies throughout the band because there are more electrons (or holes) that *just* have enough energy to cross the forbidden gap. Progressively fewer have sufficient extra energy to take them deeper into the band.

For the purposes of describing semiconductor operation, it is a good idea to have an energy-level diagram that refers specifically to the electrons and holes. Figure 6.13 looks the same as the energy-level diagrams we have used so far in this chapter, but we can use the shading to show the probability of finding electrons or holes (as charge carriers) in the bands. You will discover how useful this diagram can be for describing the way semiconductor devices work in the next chapter.

Before leaving Figure 6.13, let us be completely sure what it shows. It is an energy-level diagram for a piece of material as a whole. The upper band is the conduction band. It may be empty or it may have electrons in it. Any electrons in the conduction band can move freely through the material, and each electron carries one unit of negative electric charge. The density of the shading corresponds to the probability of finding free electrons at any given level in the band.

The space between the bands is the forbidden gap. No electrons can exist with this energy level, though they can cross the gap (apparently in zero time, although that's another story...).

fig 6.13 *an energy-level diagram illustrating distribution of holes and electrons (this diagram will be used extensively in Chapter 7)*

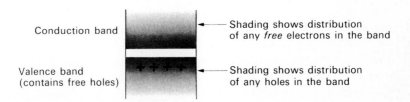

The lower band is the valence band. Its electrons are usually locked into the atomic structure, but in semiconductors there may be holes in it. Holes can drift about freely in the valence band, and are each carriers of one unit of positive electric charge. The density of the shading represents the probability of finding holes at any given level in the valence band.

We do not consider the absolute values of the energy levels, any more than we consider the inner bands of the atomic structure, for they are not relevant to the way semiconductor devices work. The energy is actually the total kinetic and potential energy of the electrons, and is a function of the physical structure of the material and also of any applied electrical potential. If an electric charge is applied to the material, the width and relative position of the energy bands will be unaltered, but the total energy in the system will change, moving the whole system of bands up and down the energy scale. Thus a *negative* electric potential applied to a material will move all the bands of the diagram *up* the energy scale, by adding to the energy (negative) of all the electrons.

An applied positive charge will move all the bands down the diagram. You will see this overall movement of energy bands in the next chapter, where we will be looking at the simplest of common semiconductor devices, the *pn* diode.

6.5 MANUFACTURE OF SEMICONDUCTORS

It is harder than you might think to manufacture a piece of *n*-type or *p*-type silicon. The first step is to purify the silicon, using the best chemical techniques available. 'Chemically pure' silicon may contain an almost negligible amount of impurity, but this is enough to make it quite useless for making semiconductor devices. Silicon in which the impurities are measured in parts per *billion* is needed. A technique known as *zone refining* was developed. Zone refining consists of taking an ingot of pure silicon, and repeatedly moving it through a radio-frequency heating coil in the same direction. Figure 6.14 illustrates this. The heated area sweeps all the impurities down to the end of the ingot, until eventually, after many passes through the coil, the major part of the bar is pure enough. The end of the ingot is cut off and sent back for chemical purification again.

Next, the silicon must be made into a crystal. Most solids, when they are cooled from a molten state, crystallise into many individual crystals, with distinct boundaries between each crystal. This is not good enough for semiconductor devices a single, large crystal is needed. The silicon is heated to a fraction above its melting point in an inert container (that is, a container made of a material like quartz, which will not react with the silicon). A single tiny silicon crystal is dipped into the molten silicon, and

fig 6.14 *zone refining a silicon ingot; the ingot is passed repeatedly through the heating coil, which sweeps all the impurities down to one end of the ingot*

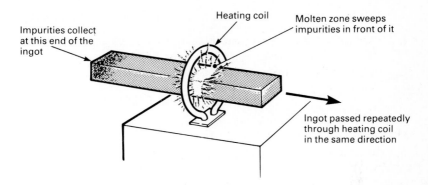

then withdrawn very slowly; the molten silicon makes the crystal 'grow', and if conditions are exactly right a sausage-shaped single crystal, about 50 mm diameter, can be drawn out.

The impurities required to make *p*-type or *n*-type silicon must also be added. There are two possible ways in which this can be done. Most obvious, the required impurities can be mixed into the molten silicon before the crystal is drawn. This method is often used. A second method, less obvious but potentially much more useful, is to add the impurities by a process known as *diffusion*. If the crystal 'sausage' is cut into thin slices—known in the trade as *wafers*—the required *dopants* (controlled impurities) can be added to each wafer as required. The wafer is first heated to about 1200°C (lower than the melting point of silicon) and then exposed to an atmosphere containing, for example, phosphorus. The phosphorus atoms diffuse into the silicon, and change it into *n*-type silicon. The usefulness of the process is that it is possible to control quite accurately the depth to which the phosphorus diffuses. This is important in making most modern semiconductor components.

The same process can be used to make *p*-type silicon, simply by replacing the atmosphere of phosphorus with one of boron.

It is even possible to add an insulating and protective layer of silicon dioxide by putting the wafer into an atmosphere of water vapour and oxygen at a high temperature. The diffusion process makes many things possible and is crucial to the manufacture of devices from diodes to integrated circuits, as we shall see in the next few chapters.

QUESTIONS

1. If an electron's energy is increased, what will happen to its orbit round the atom's nucleus?

2. Only one energy band in a atom may not be completely full. Which?

3. A conductor can carry an electric current, whereas an insulator will not. At an atomic level, what is the crucial difference between conductors and insulators?

4. Describe briefly the process used to make n-type silicon for use in transistor manufacture.

5. Explain (i) extrinsic semiconductor, (ii) forbidden gap, (iii) holes, in the context of semiconductors.

THE pn JUNCTION DIODE

The special properties of semiconductors are utilised to make a wide variety of electronic components, ranging from diodes to microprocessors. A detailed understanding of one of the simplest semiconductor devices, the *pn junction diode*, is an important prelude to a study of more complicated devices, and it is here that we begin.

A diode is basically the electrical equivalent of a one-way valve. It normally allows electric current to flow through it in one direction only. The symbol for the diode is given in Figure 7.1.

fig 7.1 *circuit symbol for a diode*

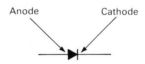

Anode   Cathode

The arrowhead in the diagram shows the permitted direction of conventional current flow (remember, electron current goes the other way). Diodes can be made to handle currents varying from microamps to hundreds of amps—a selection of types is shown in Figure 7.2.

A 'perfect' diode would probably have infinite resistance in one direction, the 'reverse' direction, and zero resistance in the forward direction. A real diode exhibits rather different characteristics, the most notable of which is a *forward voltage drop*, which is constant regardless (almost) of the current being passed by the diode. Compare Figures 7.3 and 7.4. In Figure 7.3 the diode is *reverse-biased* and does not conduct. The lamp remains out and the meter shows the whole battery voltage appearing across the diode. In Figure 7.4 the diode is *forward-biased*, the lamp lights, and the meter shows 0.7 V across the diode. By using bulbs of different wattages to pass different currents through the circuit we can show that the 0.7 V is constant. The diode used in this demonstration is a silicon

fig 7.2   *a selection of diodes of various power ratings, from 1 A to 20 A*

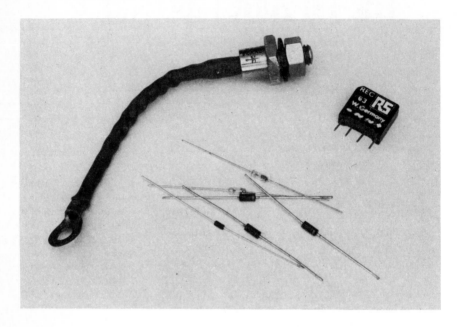

fig 7.3   *a circuit to reverse-bias a diode; note the meter is connected with its '+' to the battery '−', which results in a negative reading*

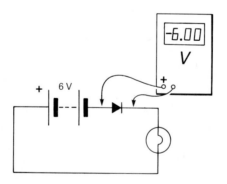

diode. If we were to replace it with one using a germanium semiconductor, the forward voltage drop would be constant at 0.3 V. What causes the voltage drop? To find out we must look at the way the diode functions at an atomic level, and go back to the energy-level diagrams.

Figure 7.5 shows an energy-level diagram for a piece of p-type silicon and a piece of n-type silicon; the two materials are not in contact.

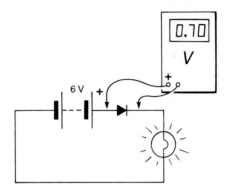

fig 7.5   *energy levels for two separate pieces of semiconductor*

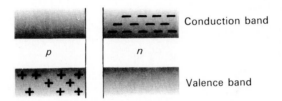

Now we place the two pieces of silicon in intimate contact (fuse them together for best results) and the energy-level diagram looks like Figure 7.6.

A few moments thought will show what has happened. The *p*-type material has extra holes and the *n*-type material has extra electrons. When the two types of semiconductor are put in contact the 'extra' electrons in the *n*-type material flow across the junction to fill up the holes in the *p*-type material. The electrons drift from the *n*-type to the *p*-type through the conduction band, and then fall down into the valence band of the *p*-type material to annihilate one hole for each electron. The holes also move, drifting from the *p*-type to the *n*-type semiconductor in the valence band.

Eventually an equilibrium is reached—but the *n*-type material has lost electrons to the *p*-type, and lost energy in the process. The *p*-type, on the other hand, has gained energy and the result is the state of affairs represented by Figure 7.6. The 'hill' in the diagram represents an area in the junction between the two semiconductor materials about 0.5 μm wide, called the *depletion region*, or sometimes the *transition region*. In this

fig 7.6 *energy levels for two pieces of semiconductor, one p-type and one n-type, in contact with each other*

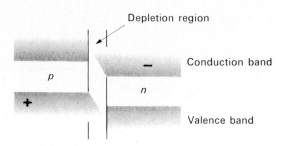

region there are no holes or free electrons. The 'height' of the hill, i.e. the difference in energy levels, can be measured in volts: for silicon it is 0.7 V, and for germanium it is 0.3 V.

The junction between the p-type and n-type semiconductor is, in fact, the active part of the pn diode. A typical low-power pn junction diode might be constructed as shown in Figure 7.7.

Details of the way the diode is manufactured can be found in Chapter 11, but for the time being it is enough to regard the diode as a block of p-type semiconductor in contact with a block of n-type semiconductor.

fig 7.7 *construction of a low-power diode (the circuit symbol underneath indicates the direction of current flow)*

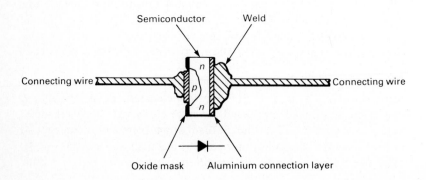

7.1 REVERSE-BIASING THE *pn* DIODE

Assume the diode is connected as in Figure 7.3, so that it is reversed-biased (non-conducting). The energy-level diagram for this state is shown in Figure 7.8.

fig 7.8 *energy levels in the reverse-biased diode*

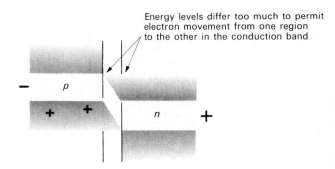

The energy of the *p*-type material is increased by the negative battery potential, while that of the *n*-type is decreased. Look at the thickness of the conduction and valence bands and at their relative positions in the two different materials, on either side of the depletion region. No parts of the conduction bands or the valence bands have the same energy levels. Electrons cannot move from the *p*-type to the *n*-type material through the conduction band because even the least energetic electrons have too much energy. Similarly, holes in the valence band of the *n*-type material all have too little energy to drift into the valence band of the *p*-type. The energy levels of the bands on either side of the depletion region are too different to allow either electrons or holes to move from one type of material to the other.

It follows that if electrons and holes are unable to cross the depletion region, there can be no current flow, for electric current is a movement of electrons (or holes) along a conductor.

7.2 FORWARD-BIASING THE *pn* DIODE

Let us reverse the situation, and connect the diode as in Figure 7.4, i.e. to forward-bias it. The energy-level diagram for this state is given in Figure 7.9. This time, the battery potential decreases the energy of the *p*-type semiconductor, and increases the energy of the *n*-type. When the potential applied is equal to or more than 0.7 V (assuming we are using a silicon diode and not a germanium one) the two types of material are at the same energy level, and the conduction and valence bands coincide. Both electrons and holes can move freely between the two types of semiconductor, and electric current can flow through the diode.

fig 7.9 *energy levels in the forward-biased diode*

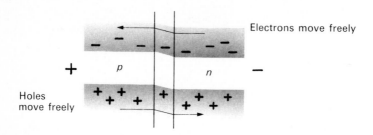

The reason for the 0.7 V forward voltage drop of the silicon diode should now be clear. A potential of 0.7 V must be applied to the *pn* junction to equalise the energy levels, and this is effectively subtracted from the potential available to push electric current through the diode.

7.3 POWER DISSIPATION IN THE DIODE

It is the forward voltage drop that sets an upper limit on the amount of current we can get through a diode. The lamp in Figure 7.4 might use, for example, 1 A, and the battery might be 12 V. In such a case a current of about 1 A flows in the circuit—but some of the power is 'wasted' in the diode as a result of the forward voltage drop, and is dissipated from the junction as heat. The amount of power that the diode dissipates can be calculated very simply using the formula:

$$P = V_f I$$

where P is the power (in watts), I is the current flowing (in amps), and V_f is the forward voltage drop of the diode in question. With 1 A flowing, the diode will dissipate 0.7×1 watts, or 0.7 W. When a diode is used in a circuit its power rating must exceed the expected dissipation, preferably by a comfortable margin.

In practice, the formula is complicated by the fact that the diode gets hot. The intrinsic conductivity of semiconductors increases with temperature (the bands get thicker) and this has the effect of lowering the forward voltage drop as the temperature rises. For both silicon and germanium the forward voltage drop decreases at the rate of about 2.5 mV/°C temperature rise.

Since the junction temperature of a silicon diode may exceed 100°C with the diode operating near its maximum current, temperature effects can become quite significant. The 'normal' forward voltage drop of 0.7 V is reduced to 0.45 V for a 100°C temperature rise. For most purposes this effect can be discounted, but in circuits where the voltage drop across the diode is important, we have to bear it in mind. It is not, incidentally, possible to reduce the forward voltage drop to zero—at very high temperatures, the junction melts. Silicon is much more tolerant of temperature than germanium. For this reason, and also because the reverse leakage current is much lower (see below), silicon diodes are the norm and germanium diodes are relatively rare—they are used only in applications where a low forward voltage drop is essential.

7.4 REVERSE LEAKAGE CURRENT

Diodes are 'imperfect' in another way. When connected in a circuit that reverse-biases it (Figure 7.3 again) a diode allows a very small current to flow. This current is called the *reverse leakage current* and is partly due to thermally generated electron–hole pairs and partly to leakage across the surface of the diode. The reverse leakage current is very small for silicon diodes, in the region of 2 to 20 nA. For almost all purposes this minute current can be ignored. The reverse leakage current for germanium diodes is much higher, typically 2 to 20 μA, and this may be important in some circuits.

The reverse leakage current is also temperature-dependent and for both silicon and germanium it roughly doubles for each 10°C rise in temperature.

7.5 REVERSE BREAKDOWN

As the voltage applied to a diode to reverse-bias it is increased, there comes a point at which *reverse breakdown* occurs. In normal *pn* diodes, this breakdown is catastrophic and destroys the diode. Depending on the construction of the diode, the point at which reverse breakdown occurs can be anywhere from fifty to a few hundred volts. Manufacturers always specify the safe reverse-bias voltage, which is normally referred to as *peak inverse voltage*. Thus a 50 p.i.V diode can be reverse-biased at 50 V without danger of breakdown taking place.

We are now in a position to draw a graph of a typical diode's conduction characteristics at room temperature (see Figure 7.10). Note that the

88

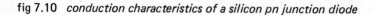

fig 7.10 *conduction characteristics of a silicon pn junction diode*

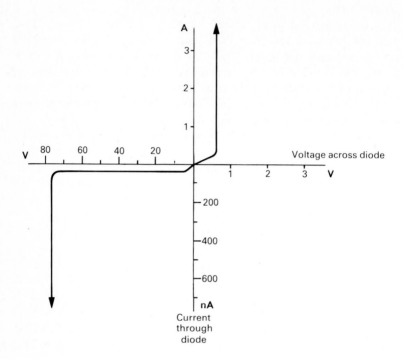

four scales of the graph are different, to show different aspects of the characteristics. On the right, the voltage across the diode is shown for different values of current through the diode. The graph continues straight up until the current reaches a value that causes overheating ($P = V_f I$) and melts the *pn* junction. On the left, the reverse leakage current is shown for increasing voltage; it remains steady at a few nanoamps until reverse breakdown occurs (at a substantially higher voltage than the 50 p.i. V manufacturer's maximum for this device) and the current increases to a value limited by factors other than the characteristics of the *pn* junction, which at this point will have just ceased to exist.

7.6 THE ZENER DIODE

There is one particular type of diode that makes use of a form of reverse breakdown to provide a constant 'reference' voltage. The circuit of Figure 7.11 shows the symbol for a Zener diode and also a suitable circuit to demonstrate its properties.

fig 7.11 *a Zener diode in a typical circuit*

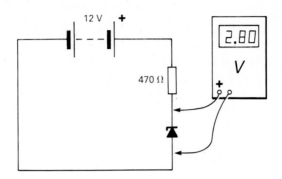

In this circuit the diode is reverse-biased (Zener diodes are always used this way) and the meter shows a drop of 2.8 V across it. The voltage drop across a reverse-biased Zener diode will be substantially constant for all values of current up to the diode's limit, and for all values of voltage higher than the Zener voltage. A resistor is necessary to limit the current through the circuit; in this case the current is calculated by Ohm's Law:

$$\frac{12 - 2.8}{470} = 19.6\,\text{mA}$$

If we were to double the battery voltage to 24 V, the current through the circuit would increase to

$$\frac{24 - 2.8}{470} = 45\,\text{mA}$$

but the voltage across the Zener diode would remain at 2.8 V. The Zener diode is therefore a valuable device for *voltage regulation*, giving a fixed reference voltage over a range of input voltages.

Zener diodes are named after C. M. Zener, who in 1934 described the breakdown mechanism involved. In fact, Zener's description applies only to diodes with a Zener voltage of less than about 3 V. Above this, a mechanism called *avalanche breakdown* begins to take over. However, both types are generally lumped together under the heading 'Zener diodes', and are available in a range of voltages from 2 to 70, and with power ratings from 500 mW to 5 W. The power dissipation of the Zener diode is calculated by means of the usual formula

$$P = V_Z I$$

where V_Z is the Zener voltage. By way of example, the Zener diode in

Figure 7.11 is dissipating

$$2.8 \times 19.6 = 54.9 \, \text{mW}$$

The conduction characteristics of a Zener diode can be drawn in a diagram similar to Figure 7.10. Look at Figure 7.10 again—the only difference in the Zener diode's characteristic is in the *reverse breakdown* (lower left quadrant), which will occur at the Zener voltage. The important factor— not shown in the diagram of characteristics—is that the Zener diode is un-damaged by reverse breakdown provided that the current through the diode, and thus its heat dissipation, is limited to a safe value.

7.7 THE VARICAP DIODE

When a diode is reverse-biased, the junction behaves like a capacitor. The p-type and n-type regions form the plates of the capacitor, and the deple-tion region acts as a dielectric. You will remember that the capacitance of a capacitor depends on the thickness of the dielectric, and it is this fact that makes the pn diode useful as a variable capacitor. If the voltage used to reverse-bias the diode is increased, the difference in the energy-levels of the p and n regions is made greater. As this happens, more of the junction thickness is depleted of charge carriers, increasing the width of the deple-tion regions. Effectively, the capacitor's dielectric has become thicker, reducing the capacitance.

A *varicap diode* is a diode in which the change of capacitance with applied reverse voltage is enhanced as far as possible by the diode's physical construction. Even so, both the overall capacitance and the amount of change are fairly small, though they are sufficient to enable the varicap diode to be used in the tuning circuits of televisions, for example. A typical varicap application circuit is shown in Figure 7.12.

A tuned resonant circuit (see Chapter 15) is formed by L_1 and C_1, VC_1 (L_1 is made variable for initial adjustments) and a variable voltage applied via R_1. If this control voltage (V_C) is changed, the capacitance of the vari-cap diode is altered, varying the resonant frequency of the tuned circuit. The control voltage could come from a TV channel-changing system, probably based on digital electronic techniques, and the resonant circuit could be used in the TV tuner.

Typical varicap diodes have a capacitance swing from 6 pF to 20 pF, for an applied reverse voltage of 2 to 20 V.

7.8 THE LIGHT-EMITTING DIODE

The light-emitting diode (LED) is an extremely useful device, and replaces miniature incandescent lamps in a whole range of applications. Almost

fig 7.12 *a varactor tuning circuit, such as might be used in the tuner of a television; changing the voltage at +V$_c$ changes the capacitance of the varicap diode and alters the resonant frequency of the circuit*

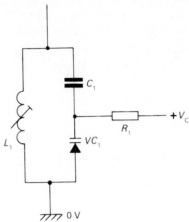

everybody is familiar with the LED display, used commonly for digital alarm clocks. LED displays are usually ruby red, but green and yellow is also available (although more expensive and therefore less common). Figure 7.13 shows two different types of LED.

Like ordinary *pn* diodes, LEDs have no inherent current limiting characteristics, and must be used with a resistor to limit the current flowing—generally to around 20 mA for a small LED indicator. A suitable circuit is shown in Figure 7.14. It is clear from the circuit that LEDs are used in the forward-biased mode: indeed, they must be protected from reverse-bias as they usually have a reverse breakdown voltage of only a few volts. If breakdown occurs, the LED is destroyed. LED displays are described in rather more detail in Chapter 16.

Mechanism of light emission
To move an electron from the valence band into the conduction band requires energy; and example of this is the creation of an electron–hole pair (see Figure 6.9 on page 74). Similarly, if an electron falls down from the conduction band to annihilate a hole in the valence band, energy is given off. In normal *pn* diodes, this energy takes the form, ultimately, of heat. In the LED, part of the energy is given off as light, the wavelength (colour) of the light depending on the distance the electron falls, which in turn depends on the width of the forbidden band. LEDs are made with a variety of rather exotic materials: gallium arsenide phosphide for red or yellow, gallium phosphide for green.

fig 7.13 *some typical-light-emitting diodes (optoelectronics is discussed in more detail in Chapter 16)*

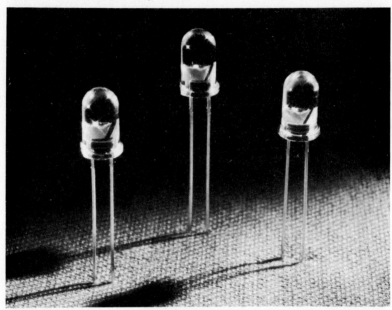

photographs by courtesy of Hewlett-Packard

fig 7.14   *a typical circuit for driving a LED; the resistor is necessary to limit the current through the LED, which is used in the forward-biased mode*

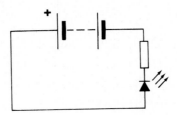

The physical design of a LED is important. Because the semiconductor material is rather opaque, most of the light produced never reaches the surface. To improve matters, the junction has to be very close to the surface, only about 1 μm or so. This means that the *p* or *n* region has to be very thin. Electrical connections to this layer have to be made in such a way that they do not obstruct the light. Figure 7.15 shows a typical LED. In practical applications LEDs are often pulsed. This is because the light output increases quite rapidly with increasing current, so a LED that is pulsed at 50 mA (and is on half the time) will look brighter than it would with a continuous 25 mA, while dissipating the same amount of heat. The pulse-repetition frequency has to be rapid enough for persistence of vision to make it look as if the LED is on continuously—more than about 50 Hz.

fig 7.15   *physical construction of a LED; light is emitted through the very thin p-type region and the device must be designed so that this light is obstructed as little as possible; a large connection area is made to the n-type region and this also serves to conduct heat away from the junction*

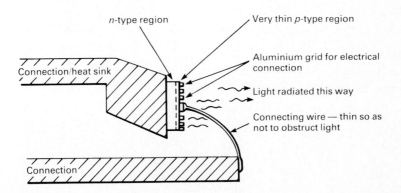

Like a *pn* diode, the LED exhibits a *forward voltage drop*. This varies according to the type and colour of the LED but is generally between 2 and 3 V.

QUESTIONS

1. What is the value for the forward voltage drop of (i) a silicon diode, (ii) a germanium diode?

2. If a silicon diode is passing 2 A, what is the amount of power that is likely to be dissipated by the diode?

3. How is the forward voltage drop of a diode affected by variations in temperature?

4. What is the main use of a Zener diode?

5. Which characteristic of semiconductor diodes is emphasised in varicap diodes?

6. A LED cannot be connected directly to a battery. What additional component is required, and why?

BIPOLAR TRANSISTORS

The transistor is perhaps the most fundamental device in modern electronics. It was the invention of the transistor that started the 'electronics revolution', and the transistor is still the most basic of all elements in an electronic circuit.

Various types of transistor are available, but there are two main classes: *bipolar transistors* and *field-effect transistors*. The first group is the subject of this chapter (the second is the subject of Chapter 9). The circuit symbol for a typical bipolar transistor is given in Figure 8.1.

fig 8.1 *circuit symbol for a transistor, npn type*

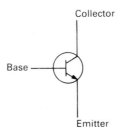

The device has three terminals, referred to as the *emitter*, *collector* and *base*. Readers who are going through this book in sequence and have read Chapter 5 will already be familiar with the thermionic triode valve, and will recognise a correspondence between the three terminals of the transistor and the cathode, anode, and grid of the valve. In many ways the transistor could be said to have 'replaced' the valve, for it is the same sort of device and is used to amplify electrical signals. But the similarity is not total. Whereas the valve responds to changes of voltage on the grid, the transistor is a *current* operated device, and provides an output current that is proportional to the input *current*. It behaves as if it were a variable resistor, the value of which depends on the current flowing through the base connection. Indeed, the name 'transistor' is a contraction of 'transfer-resistor', which goes some way to describing the properties of the device.

The circuit in Figure 8.2a shows a transistor in a test circuit configuration (but don't try it!), referred to as *common emitter* mode. If the input terminal *A* is made positive with respect to the emitter, a current will flow between the base and emitter. The base–emitter junction will behave exactly like a forward-biased semiconductor diode (see Figure 8.2b).

fig 8.2 *a simple transistor amplifier and an indication of the base circuit*

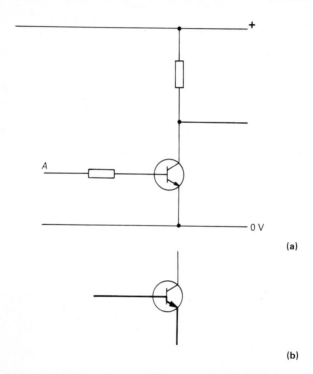

(a)

(b)

Within a certain range of *base currents*, the collector–emitter junction will exhibit the characteristics of a variable resistor, the resistance of which is inversely proportional to the base current (see Figure 8.3).

The *output current* of the transistor, measured on a meter in series with the collector, is larger than the input current by a fairly large constant, typically 20 to 1000, depending on the specific device.

A practical circuit for measuring a transistor's output current for a range of input currents is given in Figure 8.4. Adjustment of the variable resistor VR_1 permits control of the base current, and makes it possible to

fig **8.3**   *the transistor places a variable amount of resistance in the collector circuit*

fig **8.4**   *a practical circuit for approximate measurement of transistor gain*

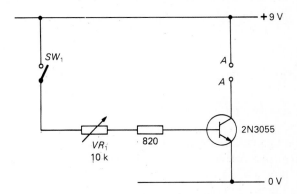

check the output current for a range of input currents. The circuit is used as follows; with SW_1 open, and VR_1 set to a maximum resistance, the meter is connected across the terminals of SW_1 to read the base current. VR_1 is set to give the required value of base current. The meter can now be removed and connected in series with the collector circuit, across terminals A-A. If SW_1 is now closed, the collector current can be measured.

The advantage of this test circuit is that only one multimeter is required; the disadvantage is that there is no compensation for the resistance of the meter, which may affect the accuracy of the experiment. If required, a resistor equal in value to the resistance of the multimeter at the current range selected may be wired in series with SW_1 for more accurate results.

Using the components specified in Figure 8.4, the test can be performed for values of collector current up to about 200 mA. The battery may begin to flag at currents in excess of this, and if the collector circuit is left connected for any length of time, the transistor will get quite warm—after all, at 200 mA (and once again assuming that the test meter has negligible resistance) it will be dissipating

$$P = V \times I \text{ watts}$$
$$P = 9 \times 200/1000$$
$$P = 1.8 \text{ watts}$$

For this reason, the specified transistor (an extremely high-power type) should be used in this circuit. Resistor R limits the base current to a safe value.

Over the range of collector currents from 10 to 200 mA, the calculation I_c/I_b (collector current/base current) should yield a value that is approximately constant. This constant is the *large-signal current gain* of the transistor when used in the common emitter mode (it can be used in other modes—more about them later in this chapter). This value is referred to by the symbol h_{FE}. The h_{FE} of the 2N3055 is quoted (by a manufacturer) as being between 20 and 70; there is quite a wide variation between different examples of the same device.

8.1 THE TRANSISTOR AMPLIFIER

It should now be clear how the transistor works as an amplifier. Figure 8.5 shows a transistor amplifier circuit at its simplest.

A voltage applied to the input will cause a base current to flow, and the base current will be reflected in a change in collector current—which in turn will alter the voltage appearing at the output. A convenient way of visualising this is to think of the transistor's collector circuit and the collector load resistor R_L as forming two halves of a potentiometer, with the lower half variable in value as the base current changes. Figure 8.6 illustrates this.

fig **8.5** *a simple transistor amplifier, unlikely to be used in practice*

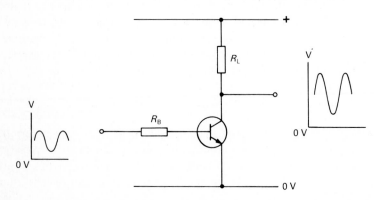

fig 8.6 *the amplifier's output circuit as a potential divider*

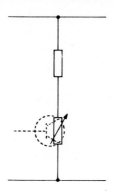

8.2 TRANSISTOR PHYSICS

Having briefly examined a typical bipolar transistor, we now turn to the physics of the device, and to the way it is constructed. A typical transistor is shown in Figure 8.7. It is a three-layer device, and is similar to a *pn* diode in general appearance, except that it has an extra layer—compare Figure 8.7 with Figure 7.7 (p. 84). An important piece of information—you will see why it is important as the description of the way the transistor operates progresses—is that the *n*-type material is more heavily doped than the *p*-type material. The result of this is that the number of free electrons in the *n*-type semiconductor greatly exceeds the number of holes in the *p*-type semiconductor.

fig 8.7 *physical construction of a silicon npn transistor*

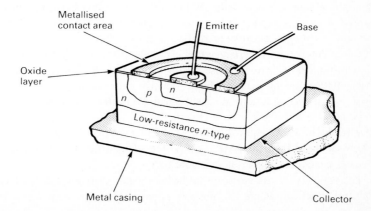

We can draw an energy-level diagram for the transistor, first when the device is 'at rest' with no power supply connected to any of the terminals. Figure 8.8 shows this.

fig **8.8** *energy-level diagram for the transistor with no voltage applied*

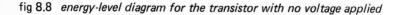

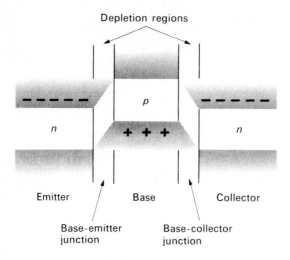

Neither electrons nor holes can cross the depletion regions, because of the large difference in energy levels. If we now make the collector positive with respect to the emitter, as in the circuit shown in Figure 8.5, the picture is different, changing to that shown in Figure 8.9. Clearly, however, there is still no flow of electrons or holes through either junction.

When the base is made rather more positive than the emitter, the state of affairs changes dramatically, as illustrated by Figure 8.10, in which a positive voltage is applied to the base to make it positive with respect to the emitter (but still negative in relation to the collector).

Under these conditions, large numbers of electrons are able to flow from the emitter region into the base region, the energy levels at the emitter-base junction having been made sufficiently close to allow electrons to move across the junction. A fairly small proportion of these electrons combine with holes in the base region. Restoration of the holes in the p-type semiconductor requires a flow of electrons out of the base region (or a flow of holes into the base region). This constitutes the base current through the circuit illustrated in Figure 8.2b.

The majority of the electrons in the base region drift towards the base-collector junction, where the positive potential attracts them across the

fig 8.9 *energy-level diagram when the collector is positive relative to the emitter*

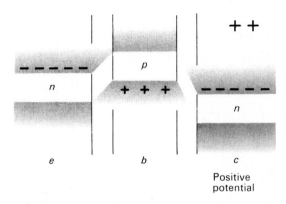

fig 8.10 *energy-level diagram when the base is more positive than the emitter*

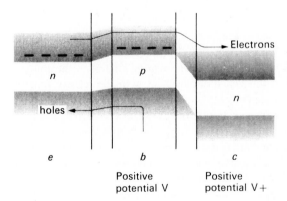

depletion region into the *n*-type material of the collector. Electrons can thus flow from the emitter to the collector, and this current flow constitutes the *collector current*, illustrated in Figure 8.3.

The collector current is larger than the base current because the relative scarcity of holes in the lightly doped *p*-type base region limits the amount of current that can be drawn from the base by restricting the availability of charge carriers.

Because both electrons and holes are involved in its operation, this type of transistor is called *bipolar*, and because of its construction, with a p-type base region and n-type emitter and collector, it is called an *npn* transistor. A few moments' thought will be enough to realise that it is equally possible to make a *pnp* transistor, having an n-type base and p-type emitter and collector. Figure 8.11 shows the circuit symbol for a *pnp* transistor, in a circuit similar to the one in Figure 8.5.

fig 8.11 *a pnp transistor used in a simple amplifier circuit*

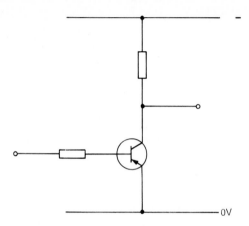

Note that the polarity of the power supply is also reversed, with the base and collector at a *negative* potential relative to the emitter. Operation is the same as the *npn* transistor, but with the relative functions of holes and electrons exchanged.

Both *npn* and *pnp* transistors are readily available; *npn* types are somewhat more common, and generally a little cheaper.

Almost all modern transistors are made using silicon as the semiconducting material. Compared with germanium (see Chapter 6) silicon is less affected by temperature variations, and has fewer undesirable characteristics. There are a few germanium transistors available, and they can be used where a low forward voltage drop is essential.

8.3 BIPOLAR TRANSISTOR CHARACTERISTICS

The circuit configuration shown in Figures 8.5 and 8.11 is referred to as a *common emitter* circuit, the emitter being common to both input and output circuits. It is the most usual way of connecting a transistor. When used in this (or indeed, any other) mode, the transistor exhibits certain characteristics which we can use as a measure of its performance.

First, and perhaps most important, is the *large signal current gain*, the value we measured at the beginning of this chapter, expressed as the ratio I_c/I_b, collector current divided by base current, and given the symbol h_{FE}. Unfortunately for those of us who like life to be simple, h_{FE} is not constant, but varies with collector voltage. This is known as the *Early effect* (after the man who first suggested what caused it). One of the best ways of illustrating the way h_{FE} changes with collector voltage is to plot a graph of collector current (I_c) against collector voltage (V_{ce}) for a range of different base currents (I_b), choosing values for I_b and V_{ce} that are sensible for the transistor being measured. A typical set of graphs—called *characteristic curves*—is shown in Figure 8.12, which is worth careful study.

fig 8.12 *transistor-characteristic curves*

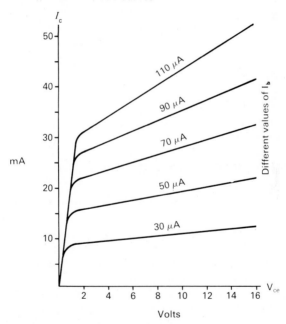

The straight, sloping part of the curve is called the *linear region*, and the transistor should be operated on this part of the curve for most purposes, for example for all types of linear amplifier. The almost vertical part of the curve marks the *saturated region* of the characteristics—the transistor 'switches' rapidly from a non-conducting to a conducting state (from 'off' to 'hard on') as the base current is increased. Transistors are operated in this part of the characteristic in digital circuits, where rapid switching and low power dissipation are important.

A feature of all transistors is *leakage current*, the current that flows into the collector when the base connection is disconnected (i.e. open-circuited). The base leakage current is given the symbol I_{ceo}. For silicon transistors I_{ceo} is very small, seldom as much as a few microamps. Germanium transistors have a much larger value of I_{ceo}, often approaching a milliamp; in germanium transistors the leakage current is also intensely temperature-dependent—so much so that they can be used as temperature-sensing devices. Silicon transistor leakage current is less affected by temperature at normal operating temperatures.

Temperature changes also affect the base–emitter junction voltage drop. For each $10°C$ rise in temperature the base–emitter voltage is reduced by about 20 mV. Increasing temperature also changes the characteristic curves, spreading them out further apart and moving them upwards on the graph.

Finally, an increase in ambient temperature reduces the power-handling ability of a transistor by slowing the rate of heat transfer away from the casing (and thus the junction). Heat transfer from the transistor junction to the outside world is proportional to the temperature difference between them; a transistor's ability to dissipate heat is quoted in $°C/W$, being the temperature rise of the case (not the junction) per watt of dissipation at a given ambient temperature, usually $20°C$. *Heat radiators or heat sinks* can be used to improve the heat-transfer capability by a substantial amount— Figure 8.13 shows some typical forms.

8.4 SPECIFYING BIPOLAR TRANSISTORS

There are many thousands of different types of transistor, with every possible different specification. Manufacturers are continuously introducing new designs and withdrawing obsolete types. Broadly, there are 'design' types (the newest, for new equipment), 'stock' types and 'service' (obsolete) types. When a designer specifies a particular kind of transistor, the following main parameters are taken into account:

(1) **Class**. There are at least three classes of semiconducting device, according to its reliability, capacity for withstanding unpleasant environments—e.g. heat, damp, radiation—and physical construction. One large semiconductor manufacturer calls these classes 'military', 'industrial' and 'entertainments'. The best—military—components are substantially more expensive than the other two classes.

(2) **Power**. There are big transistors and small transistors. The big ones can dissipate as much as 150 W, whereas the smallest may overheat above 100 mW. Transistors with large dissipations are called *power transistors* and are often designed so that they can be bolted to a suitable heat radiator. The dissipation is partly a function of the size of the semiconductor chip,

fig 8.13   *a selection of heat radiators and heat sinks*

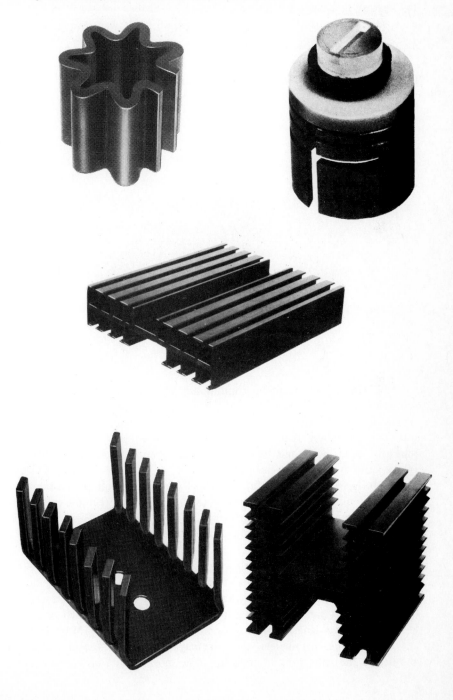

and partly a function of the arrangements inside the transistor for conducting heat away from the junction. A power transistor and a small plastic encapsulated transistor are compared in Figure 8.14. Power dissipation is given the symbol P_{TOT}.

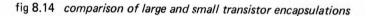

fig 8.14 *comparison of large and small transistor encapsulations*

(3) **Current**. The maximum continuous collector current a transistor can pass is given the symbol $I_{C(MAX)}$. It is related to power in that power transistors usually have a larger value of $I_{C(MAX)}$ than small-sign transistors, but the power is of course a factor of current *and* voltage. Most transistors can withstand collector currents several times larger than $I_{C(MAX)}$ if they are of very short duration.

(4) **Voltage**. The most useful measure of the working voltage of a transistor is the maximum safe voltage that can be applied between the collector and emitter when the base connection is open-circuited. This is given the symbol V_{ceo}, and may be anywhere between ten and a few hundred volts.

(5) **Gain**. We have already dealt with h_{FE}, the most convenient measure of a transistor's current gain. There are other ways of measuring the gain (in common-base mode, for instance), but h_{FE} is the most commonly quoted and used parameter. The gain may be measured for a small signal instead of a large signal, in which case the symbol h_{fe}, is used. More than any of the other parameters, the gain may vary from one example to another of the same type of transistor. The most minute differences in

the chemical compositions of the semiconductor and dopants, and tiny differences in manufacture, can make a substantial difference to the gain of a particular device. Consequently, manufacturers quote a 'typical' gain for a transistor type, and often a maximum and minimum. A variation of ±50 per cent is quite usual.

(6) Frequency. A transistor's ability to 'follow' high frequency signals depends on many factors, but principally on the width of the base region: the narrower the base region, the quicker electrons can cross it. Small signal transistors (*npn* silicon planar) would typically have a maximum usable frequency limit—given the symbol f_T—of 50 to 300 MHz. Power transistors, on the other hand, have much lower frequency limits, up to 100 MHz being typical for a 10 W type. At present the maximum f_T available, outside very expensive specialist devices, is around 1 GHz; but the power would be very low.

(7) Case. There are many different case designs, but most transistors use one of a relatively few shapes that have become 'standard' over the past few years. For small signal transistors, the T018 metal encapsulation or the T092 plastic (cheaper) are commonly used. In these, as in all transistor case designs, the shape of the case and the position of the leads indicates which of the three wires is which—they are *not* usually marked 'c', 'b' and 'e'. For transistors having a P_{TOT} of more than 500 mW or so, a larger metal case is used, generally the T05 design. For power transistors, a plastic T0126 or T0202 encapsulation with a metal tag for bolting to a heat sink or heat radiatior might be used; and for the larger power transistors (5-150 W) the T03 case is usual—this one is unequivocally intended to be used bolted to a heat sink.

All six of these common case types are shown in Figure 8.15.

fig 8.15 *six common types of transistor encapsulation*

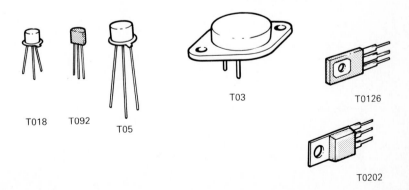

fig 8.16   *table of some commonly used transistor types*

Selection of common transistor types

Purpose	Type number	Encapsulation	Type†	h_{FE}	V_{CEO}	$I_{C(MAX)}$	$f_{T(MAX)}$	P_{TOT}
Small signal	BC107	TO18	npn Si	110–450	45 V	100 mA	250 MHz	360 mW
Small signal	BC109	TO18	npn Si	200–800	20 V	100 mA	250 MHz	360 mW
Small signal (complement: BC109)	BC479	TO18	pnp Si	110–800	40 V	150 mA	150 MHz	360 mW
Medium power general purpose	BFY51	TO5	npn Si	40	30 V	1 A	50 MHz	800 mW
General purpose	2N3702	TO92	pnp Si	180	25 V	300 mA	100 MHz	300 mW
Medium power	BFX88	TO5	pnp Si	100–300	40 V	600 mA	100 MHz	800 mW
General purpose	OC71*	TO1	pnp Ge	30	30 V	10 mA	5 kHz	125 mW
High power	2N3055	TO3	npn Si	20–70	60 V	15 A	1 MHz	115 W

* Obsolete device (included for comparison purposes).
† Si = silicon; Ge = germanium.

8.5 SOME TYPICAL TRANSISTORS

Figure 8.16 gives a table of some common transistors likely to be found in any good electronic component shop.

Next, we will look at the second major class of transistors, the *field-effect transistor* (FET).

QUESTIONS

1. Which terminal of a bipolar transistor is common to both input and output circuits in the most commonly used configuration?

2. How do you calculate the current gain of a transistor?

3. Should the emitter be more positive or more negative than the collector in (i) *pnp*, (ii) *npn* transistors?

4. Describe the *Early effect*.

5. What is the *leakage current*?

CHAPTER 9

FIELD-EFFECT TRANSISTORS

Field-effect transistors are a more recent development than bipolar transistors, and make use of a completely different mechanism to achieve amplification of a signal. Field-effect transistors (FETs) are *unipolar*, and involve only one type of charge carrier (electrons or holes) in their operation. There are also two major types of FET, junction-gate field-effect transistors (JUGFETs) and insulated-gate field-effect transistors (IGFETs). There are, as we shall see, subdivisions within these two classes.

Operationally, FETs are more similar to valves than are bipolar transistors. The main distinguishing characteristic compared with bipolar transistors is the fact that they are *voltage-controlled* rather than current-controlled. The circuit symbol is given in Figure 9.1; a voltage applied to the gate is varied to provide a corresponding charge in the resistance between the *source* and *drain*.

fig **9.1** *circuit symbol for an n-channel field-effect transistor*

Unlike the bipolar transistor's base connection, the gate of the FET has a very high input resistance, at least a few tens of megohms and in some cases, gigohms (see Figure 2.6, p. 12). The amount of current drawn by the gate is therefore extremely small.

FETs can be used in amplifier circuits, just like bipolar transistors. Compare the circuit of Figure 9.2, which shows a typical JUGFET, with that in Figure 8.5 (p. 98). The difference is the lack of a gate (base) resistor; because negligible current flows in the gate connection, such a resistor would make no difference to the operation of the circuit, adding a moderate amount of resistance to one that is extremely large in the first place. Just as there are *n*-type and *p*-type resistors, so there are *n*-channel and *p*-channel JUGFETs. Figure 9.3 shows a *p*-channel version, while those in Figures 9.1 and 9.2 are *n*-channel. As in the case of bipolar transistors, practical operation is the same but with all polarities reversed.

fig 9.2   *a simple amplifier using an n-channel FET*

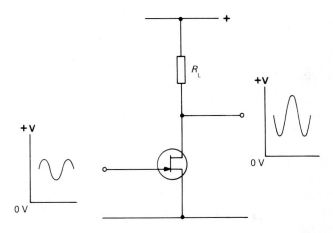

fig 9.3   *a simple amplifier using a p-channel FET (note the different polarity of the power supply, compared with Figure 9.2)*

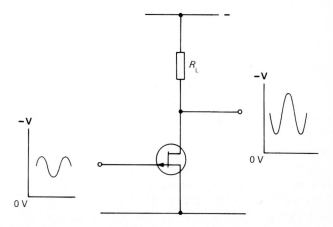

9.1 FET PHYSICS

The JUGFET has a physical structure that can be represented by a diagram like the one in Figure 9.4, though in practice it is not easy to diffuse impurities with *both* sides of the wafer, and a rather different layout is used. Chapter 11 gives details.

fig 9.4   *theoretical construction of an n-channel depletion-mode FET*

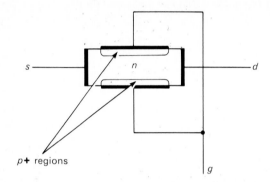

A bar of *n*-type semiconductor (almost invariably silicon) is made with shallow *p*-type regions in the upper and lower surfaces. These are connected to the gate terminal, and the two ends of the bar are connected to the source and drain.

If the bar is connected to a voltage source, current will flow through it. Since the bar is symmetrical, it can flow either way, the source and drain being interchangeable. The current flow consists of electrons moving through the *n*-type semiconductor (Figure 9.5).

fig 9.5   *movement of electrons with no voltage applied to gate*

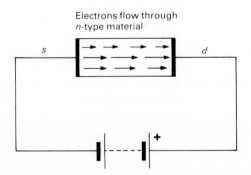

Now observe the effect of a negative potential applied to the gate regions. The junction between the p and n regions forms a reverse-biased diode (see Chapter 7) so no current flows, but an electric field extends into the n-type bar from the p-type regions. This charge forces current carriers (electrons) away from the region, reducing the amount of bar available for conducting the current between the source and drain (shown diagrammatically in Figure 9.6).

fig **9.6**   *the electric field forces charge carriers away from the plates when a negative potential is applied to the gate*

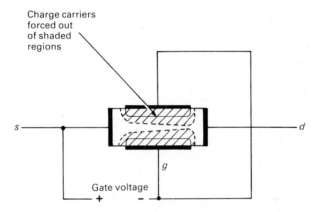

If the potential applied to the gate is made sufficiently negative, the electric field will extend across the whole thickness of the bar of n-type semiconductor, hardly any charge carriers will be available for current flow, and the current available from the drain will drop to a very low value (never to zero, for it is physically impossible for the channel to 'close' completely: see Figure 9.7).

Changes in the voltage applied to the gate will cause corresponding changes in the current flowing between the source and drain, which makes the operation of the FET very similar to that of a bipolar transistor.

9.2 THE INSULATED-GATE FET

Generally known as a MOSFET (metal-oxide semiconductor FET), the insulated-gate FET is one of the most important devices in the electronics industry. There are two basic categories of MOSFET, known as *depletion* MOSFETs and *enhancement* MOSFETs. They work on a different principle from the JUGFETs that we have been looking at so far in this chapter.

fig 9.7   *although the current flow can be greatly reduced, it is impossible to stop it completely—there will always be a small gap between the two areas without charge carriers*

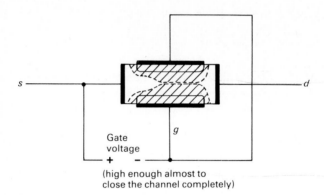

Gate
voltage
+   −

(high enough almost to
close the channel completely)

The structure of a *p*-channel enhancement MOSFET—again, a theoretical structure—is shown in Figure 9.8.

The most striking feature is the gate. It is insulated from the silicon by a thin layer of silicon dioxide. The layer is *very* thin, typically only about 0.1 μm thick. Although very thin, the silicon dioxide layer has an extremely high resistance, so the gate input resistance is very high indeed, at least 10 GΩ.

The *n*-type silicon has two regions of heavily doped *p*-type impurity, connected to the source and drain. With no applied gate voltage, one of the

fig 9.8   *theoretical construction of an n-channel enhancement mode FET (note that the gate is completely insulated from the rest of the structure by a very thin layer of silicon dioxide)*

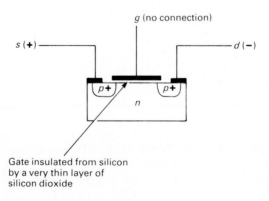

g (no connection)

s (+)   *d* (−)

p+   *p+*

n

Gate insulated from silicon
by a very thin layer of
silicon dioxide

p-n junctions (depending on which way round the source and drain have been made) will act like a reverse-biased diode and block any flow of current. If a negative potential is applied to the gate electrode, holes from the *p*-type regions are attracted into the area immediately beneath the electrode. This effectively, if temporarily, makes a narrow *p*-type region just beneath the gate, illustrated in Figure 9.9.

fig 9.9 *a p-type region is induced in the n-type bar when the gate is made negative*

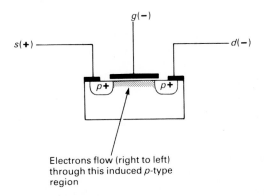

Electrons flow (right to left)
through this induced *p*-type
region

The blocking *p–n* junction is bypassed by this induced channel of *p*-type material, and electrons can flow through the device, between the source and drain. The circuit symbol for the *p*-channel enhancement MOSFET is given in Figure 9.10. The centre connection in the circuit symbol (with the arrowhead) is a connection to the silicon substrate (the chip itself). Often the connection is made internally to the source, but sometimes manufacturers fit a fourth lead so that the substrate can be used as a second 'gate', the conduction characteristics of the device then depending approximately on the difference in potential between the gate and substrate.

fig 9.10 *circuit symbol for a p-channel enhancement MOSFET*

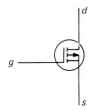

When the circuit symbol has an arrowhead pointing away from the gate, it symbolises a *p*-channel device. It is, however, equally possible to make an *n*-channel device, with all the polarities reversed, using *n*-regions diffused into a *p*-type bar. In this case the charge carriers are electrons attracted from the *n*-regions; otherwise operation is the same. The circuit symbol for an *n*-channel enhancement MOSFET is given in Figure 9.11.

fig 9.11 *circuit symbol for an n-channel enhancement MOSFET*

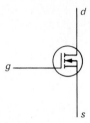

The second class of MOSFET is the depletion MOSFET. The structure is shown in Figure 9.12, and is similar to that of the enhancement MOSFET. Notice that there is an 'extra' region. A narrow strip of *p*-type impurity has been diffused into the space below the gate, so that the depletion MOSFET, with *no* signal applied to the gate, looks rather like the enhancement MOSFET when its gate is connected to make it conduct. Compare Figures 9.12 and 9.9.

fig 9.12 *theoretical structure of a depletion MOSFET*

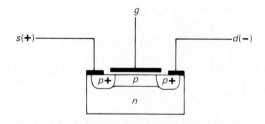

Applying a positive signal to the gate causes electrons from the *n*-type region to be attracted to the area under the gate electrode, neutralising some of the holes in the *p*-type channel and reducing the amount of current flowing between the source and drain. The higher the positive potential (for a *p*-channel device, of course), the more the source–drain current is cut off.

Actually the depletion MOSFET can also be used in the enhancement mode as well. Applying a *negative* voltage to the gate of a *p*-channel device will increase the source–drain current by adding to the number of holes available as charge carriers. Figure 9.13 shows the circuit symbols for *p*-channel and *n*-channel depletion MOSFETs.

fig **9.13**   *circuit symbols for (a) p-channel and (b) n-channel depletion MOSFETs*

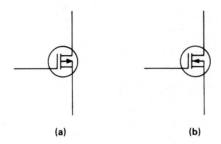

(a)   (b)

9.3 USES OF MOSFETs

At first it seems paradoxical that, even though the MOSFET is a vitally important device, individual MOSFETs are rarely used. Only in the highest-quality communications receivers do we find MOSFETs used as a matter of common practice. MOSFET technology is used overwhelmingly in integrated circuits.

Before leaving FETs for the time being, let us just look at one or two interesting aspects of the devices:

(1) Because of the necessity for a very narrow channel, it is difficult to make FETs that will carry high currents. Most FETs are therefore low-power devices.

(2) Very high frequency operation of FETs is hampered by the internal capacitance, a byproduct of the very narrow regions. There is an effective capacitance of a few picofarads between the source (and drain) and the gate.

(3) It is possible to make FETs very small.

(4) Because of the very high input resistance and the thinness of the silicon dioxide insulating layer, MOSFETs are very liable to damage from high voltages accidentally applied between the gate and the other terminals. For this reason, most MOSFETs are made with a Zener diode connected between the gate and substrate. Normally, the Zener diode is non-conducting, but if the gate voltage rises too high, it conducts and discharges the gate. Even so, MOSFETs are easily damaged by electrostatic voltages. The

charge on a person wearing shoes with rubber or plastic soles can rise, on a dry day, to several kilovolts. This is often enough to destroy the MOSFET, Zener and all. Even *touching* the gate terminal can therefore destroy the device! The terminals of a MOSFET are generally shorted together with metal foil or conductive foam until it is installed in a circuit. Extra care is needed when servicing any equipment that might include MOSFETs.

QUESTIONS

1. What is the major difference in characteristics between IGFETS and JUGFETs?

2. Special precautions are needed when handling IGFETS—why?

3. Why are high-power FETs uncommon?

4. Draw the circuit symbols for (i) *p*-channel enhancement FET, (ii) *n*-channel enhancement FET, (iii) *p*-channel depletion FET, (iv) *n*-channel depletion FET.

5. What limits the higher frequency performance of FETs?

AMPLIFIERS AND

OSCILLATORS

The last two chapters showed how transistors–bipolar and FET–work as amplifying devices, using a small current or voltage to control a much larger current. The application to a machine like a record-player, for example, is an obvious one, for the small electric signal produced by the player's pick-up cartridge must be amplified to a sufficiently large extent to drive a speaker. A system diagram of a record-player amplifier looks quite simple (see Figure 10.1). The symbols for the cartridge, amplifier and speaker are those conventionally used. In practice the amplifier is rather complex, and breaks down into two main sections: the *preamplifier*, which deals with the amplification of the small signal from the cartridge; and the *power amplifier*, which deals with the high-power amplification necessary to drive the speaker. *Audio amplifiers* are covered in more detail in Chapter 13, but in this chapter we shall look at typical techniques of *small-signal amplification*.

fig 10.1 *system diagram for a record-player amplifier*

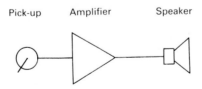

Pick-up   Amplifier   Speaker

At its simplest a transistor amplifier circuit looks like the one in Figure 10.2, but although this simple configuration is satisfactory for demonstrating 'transistor action', it is incapable of amplifying an audio signal.

Consider the effect of applying the alternating voltage from the pick-up cartridge to terminal A as shown. Since the transistor conducts only when the base–emitter junction is forward-biased, only the parts of the signal that are positive relative to the emitter will cause the transistor to conduct.

fig 10.2 *the simplest form of transistor amplifier; this circuit could not amplify an audio signal!*

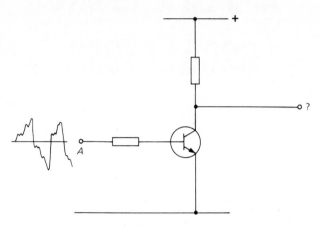

As well as amplifying the input signal, the transistor is rectifying it. Occasionally this happens in an amplifier under fault conditions, and it sounds terrible! It is also clear that a proportion of the positive part of the signal is lost as well, since the transistor will not conduct, even with the base-emitter junction forward-biased, until the potential exceeds 0.7 V for a silicon transistor. This is a substantially higher voltage than a magnetic pick-up cartridge produces, so in practice the output of the circuit in Figure 10.2 would be zero.

A solution to the problem is simply to connect a suitable positive potential to the base, ensuring that it is always forward-biased. This is best done with a potential divider network involving two resistors, as shown in Figure 10.3.

This does at least produce an output, if the two bias resistors R_1 and R_2 are exactly the right values; the resistors should be chosen so that the collector is at half the line voltage with no signal applied. This gives approximately equal 'headroom' for the signal in its positive and negative excursions. Referring to the load lines in Figure 8.12 (p. 103), it means that the *operating point*—that is, the mid-point of the variations in collector voltage when the amplifier is working—is halfway along the load line.

You should recall from Chapter 8 that because of tiny differences in manufacture it is not possible to specify the gain of a transistor very accurately, a gain variation of ± 50 per cent being common. Using the circuit in Figure 10.3 would mean careful measurements and a new pair of resistors for each individual transistor used—not an ideal requirement for mass production. The leakage current of the transistor, and thus its operat-

fig 10.3 *a bias network for a transistor, enabling the transistor to be used with audio signals (the resistor values would need to be altered for each individual transistor)*

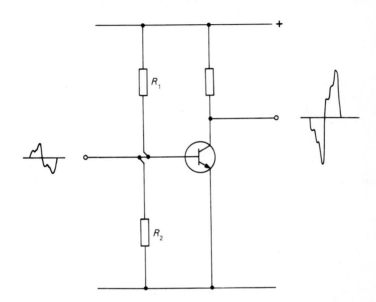

ing point, will also change with temperature. What is needed is a circuit that will automatically compensate for variations in gain of the transistor and ambient temperature. Such a circuit is shown in Figure 10.4.

In this circuit it is the values of the resistors that determine the operating point. The gain of the transistor, leakage and temperature have practically no effect. The circuit involves a *feedback* loop. Feedback is a principle that is often used in electronics, and you will find it in many different contexts. In this case *negative feedback* is used. Negative feedback consists of a connection from the output of a system back to the input, arranged so that the change in output reduces whatever is causing that change. Negative feedback improves the stability of a system, in that any change is counteracted.

The opposite, *positive feedback*, makes a system less stable, as the effect of any change is augmented by the feedback.

Although the circuit of Figure 10.4 looks simple, it is not immediately obvious how it works. Its operation is as follows. First, the base of the transistor is held at about +1.5 V, high enough to overcome the 0.7 V potential barrier of the transistor's base–emitter junction if the collector is near 0 V. Consider what happens if the transistor were non-conducting; the emitter is connected to 0 V via the 1.5 kΩ resistor, and so is at 0 V.

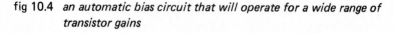

fig 10.4 *an automatic bias circuit that will operate for a wide range of transistor gains*

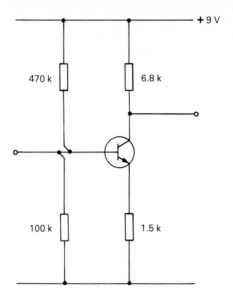

However, this means that there is a potential difference of about 1.5 V across the base-emitter junction, so in practice the transistor is turned on and conducts.

A collector current is therefore flowing, and the collector resistor, transistor, and emitter resistor operate as a potential divider. Clearly the potential on the emitter will be higher than 0 V. Neglecting the voltage drop in the transistor, the emitter will be at a maximum potential of 1.5 V if the transistor is turned 'hard on'—but, of course, if the base and emitter are at roughly the same potential, there will be no base current, and the transistor will be off!

In practice, the base current settles down to a fixed value, and in the circuit shown the collector will be 'balanced' at about half the supply voltage.

The compensating action of the circuit is easy to see. Assume that an increase in temperature causes the leakage current to rise and the transistor's collector-emitter resistance to drop. The resistance in the upper half of the potential divider (of which the transistor's emitter is the middle) will be lowered, and the positive potential on the emitter will rise, to become closer to that of the base. This will reduce the base current, and compensate for the temperature change. Transistors with different values of h_{FE} are also compensated for in the same way.

A simpler but somewhat less effective bias circuit is shown in Figure

10.5. In this design the bias is fed to the base via a resistor connected between the collector and base. Any tendency for the transistor to conduct more will reduce the positive potential at the collector, since the transistor and the 10kΩ load (collector) resistor form the two halves of potential divider. Ohm's Law indicates that this reduces the current flow through the 100 kΩ bias resistor, since the voltage across it is reduced. Once again, we have a circuit in which any increase in the transistor's collector current causes a drop in base current.

fig **10.5** *a simpler circuit than that of Figure 10.4, but sufficient to provide automatic bias for silicon transistors in many applications*

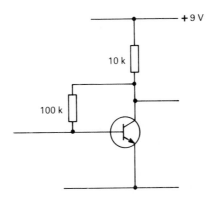

10.1 CAPACITOR COUPLING AND BYPASS CAPACITORS

When an amplifier is used to amplify an audio waveform, or indeed any other continuously changing quantity, the amplifier of Figure 10.4 is connected to the input device (a pick-up cartridge was mentioned) through a capacitor. The purpose of this is to block any d.c. component in the input. Figure 10.6 illustrates an amplifier circuit connected to a magnetic pick-up.

The pick-up provides a signal source, and this is *coupled* to the base of the transistor with a capacitor. The capacitor blocks any flow of d.c. which would otherwise interfere with the bias arrangements. If the capacitor were omitted, the resistance of the coil in the pick-up, typically just a few ohms, would be in parallel with the lower resistor in the base bias system. This would reduce the potential on the base almost to 0 V, which would turn the transistor off. For much the same reasons, the output of the amplifier would be coupled to the next stage of amplification with a capacitor.

fig 10.6 *a pick-up preamplifier; small capacitors couple the input and output of the amplifier, and a large electrolytic capacitor is used to bypass the emitter resistor, which improves the efficiency of the amplifier without altering the d.c. bias conditions*

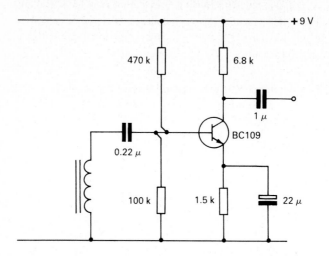

fig 10.7 *a two-stage amplifier, using the same circuit configuration as Figure 10.6*

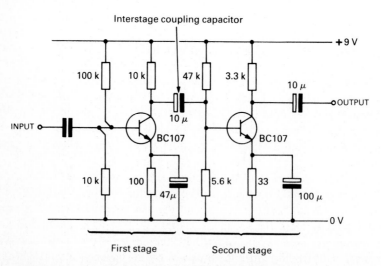

Additionally, the emitter resistor reduces the efficiency of the amplifier by allowing the potential of the emitter to change with the signal being amplified. A large-value capacitor, connected across the emitter resistor, prevents the signal component affecting the bias conditions. An electrolytic capacitor is generally used for the emitter resistor bypass capacitor. Figure 10.7 illustrates a two-stage amplifier based on the circuit in Figure 10.6, with capacitor coupling between stages. Note the difference in resistor values for each stage—the second stage has a larger input voltage swing than the first, and also delivers a higher output current.

10.2 GAIN OF MULTISTAGE AMPLIFIERS

When calculating the total gain of amplifiers having more than one stage, it is important to realise that the total gain of the system is equal to the gains of each individual stage *multiplied* together, not added. If the stage gain in a two-stage amplifier were 30, then the total gain of the system would be 30 × 30 = 900.

10.3 NEGATIVE FEEDBACK

Feedback is often applied over more than one stage. Figure 10.8 shows a commercially produced design incorporating this feature.

fig **10.8** *a commercially produced amplifier, intended for use with a tape-player; the input is provided by a tape playback head, and the output is fed to an audio power amplifier (see Chapter 13)*

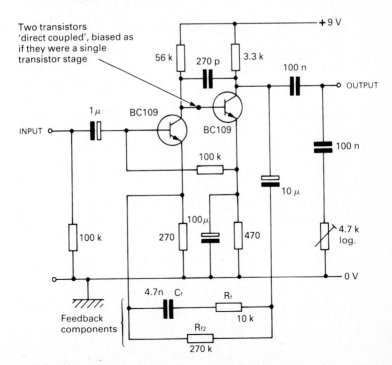

It is possible to use feedback to alter the frequency characteristics of an amplifier. It is often necessary to limit the high frequency amplification of a system, and this is easily done using a feedback capacitor instead of a resistor. The capacitor's use as a 'frequency sensitive resitor' is clear. At low frequencies, the small value of capacitance feeds back only a tiny proportion of the signal, whereas a progressively larger proportion of the signal is fed back as the signal frequency increases.

In Figure 10.8 the feedback line is from the collector of the second transistor to the emitter of the first transistor. It is coupled via a large capacitor $(10\,\mu F)$ so that it does not affect the bias conditions. Frequency-selective components in the feedback line establish the frequency response of the amplifier: C_f, R_f and R_{f2} are calculated so that this amplifier gives the right frequency response for a tape playback head. The amplifier was designed for a high-quality tape-player.

The 270 pF capacitor between the two transistor collectors operates to limit the high frequency response of the amplifier, and the variable resistor

fig 10.9 *another tape-player amplifier, using slightly different circuit techniques; the transistors are directly coupled*

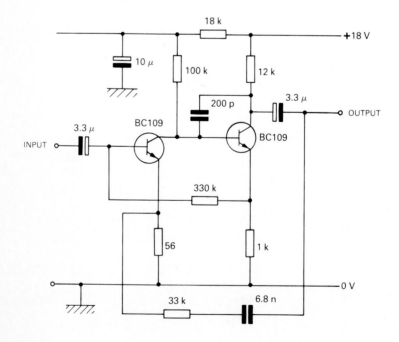

fig 10.10 *a third tape-player preamplifier design—there are many varia-*
tions on this theme (this circuit is from a commercially pro-
duced tape player)

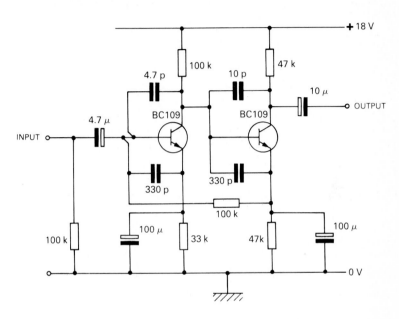

across the output, in series with a capacitor, trims the overall frequency to suit the amplifier that follows.

There are many variations in amplifier design, though there is a basic similarity between types, as one might expect. Figures 10.9 and 10.10 illustrate two more commercially produced designs—another two tape-player preamplifiers.

10.4 TRANSFORMER COUPLING

Transistor stages are generally coupled with capacitors, as in the preceding three circuits, but they can also be coupled with transformers. Originally this was the preferred method, but as components and circuit design have improved transformers have been used less and less.

The resistance of the transformer's primary winding can form the collector load of the first stage of a two-stage amplifier. Transformer coupling is illustrated in Figure 10.11. Transformer coupling is not widely used now, mainly because of the size (at least 2 cm cube) and the relatively high cost of the transformer.

fig 10.11 *a two-stage amplifier using transformer coupling between stages and at the output (transformer coupling is now little used)*

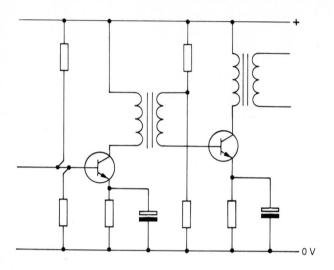

10.5 STAGE DECOUPLING

The circuit in Figure 10.7 may require a couple of extra components to make it work properly in conjunction with a *power amplifier* stage. Consider an amplifier that is built into a tape player. The output from a cassette tape head is very small—a few hundred microvolts—but the player's power output into the speaker may be as much as a watt, even for a small portable machine. The amount of amplification is therefore considerable. Power to drive the speaker has to come from batteries, and under heavy loads the battery voltage will drop. Sudden loud noises on the tape will cause rapid voltage drops.

Inevitably, the voltage changes will be reflected by changes in the low-power amplifier's quiescent output voltage, which is stabilised at about half the supply voltage. The variation in voltage of the first stage is duly amplified by the second stage, and a large and completely unwanted signal propagates through the system. The result is *instability*—the amplifier oscillates at a low frequency, often with a characteristic sound called, for obvious reasons, 'motor-boating'.

The cure is to prevent the rapid changes in voltage supplied to the first stage of amplification. This, fortunately, is easily arranged. The circuit is shown in Figure 10.12. Capacitor C_d charges up through R_d until the voltage across it is almost equal to the supply voltage. The values of C_d

and R_d are typical for a small-signal audio amplifier. Because the first amplifier stage is concerned with small signals and has small power requiremens, R_d does not have to pass a large current and the voltage dropped across R_d is small—only about 0.5 V with a 1 mA drain.

fig 10.12 *power-supply decoupling*

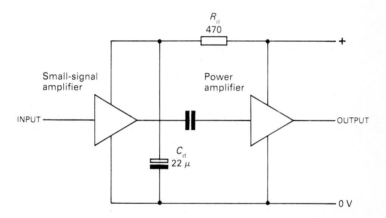

Imagine a powerful transient—perhaps a bass drum—causing a momentary voltage drop of 1 V in the power supply to the system. Neglecting for a moment the current drain of the first stage, C_d has about 0.5 V higher potential across it than the supply voltage, but cannot discharge immediately. It can only do so via R_d, at an *initial* rate of about 1 mA (by Ohm's Law), which rapidly decreases as the potential by which the capacitor exceeds the supply voltage diminishes. Adding the amplifier's current drain, it is nevertheless quite a long time—in electronics terms—before the supply to the first stage drops enough to affect the output.

Protected against sudden changes in supply voltage to the low-power stages, an amplifier is much more stable. Power-supply decoupling is a feature of almost all amplifier designs which have a large power output and large overall gain. Exceptions to this are to be found in *integrated circuit amplifiers*; the need for stabilisation of the power fed to the early stages still exists, but is fulfilled by more sophisticated *voltage regulator* systems.

10.6 INPUT IMPEDANCE

The input impedance of an amplifier is the ratio of the input voltage to input current—it is expressed in ohms, and can be thought of a being similar to resistance, but applicable to alternating as well as direct voltages

and currents. The input impedance of an amplifier is an important parameter, and is considered when the amplifier is connected to an external
signal source—such as a record-player pick-up or tape head.

A cheap record-player might well use a *crystal* pick-up cartridge (see
Chapter 13) which generates a relatively high voltage output signal—as
much as a volt—but has a very high internal resistance. Figure 10.13*a*
shows the cartridge connected to an amplifier. If the cartridge has an
internal resistance of 5 MΩ—a typical value—and the amplifier has an input
impedance of 2500 Ω, then the total current flowing in the input circuit,
Figure 10.13*b* will be

$$I = V/R$$
$$I = 1/5002500$$
$$I = 0.2 \, \mu A$$

and of the total 1 V, only 2500/5 000 000, or about 0.5 mV, will be available to the amplifier.

fig **10.13**

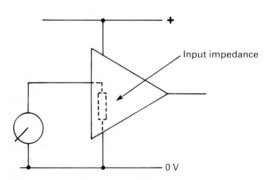

(a) *a record-player pick-up, connected to an amplifier; the
impedance 'seen' by the pick-up affects the efficiency
with which power is transferred from the pick-up to the
amplifier*

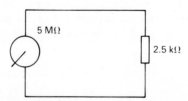

(b) *a high impedance (crystal) pick-up feeding into a low
impedance amplifier is not very efficient*

An amplifier with a higher input impedance should therefore be used, to maximise the efficiency and make best use of the high output voltage of the crystal pick-up. Conveniently, a FET can be used, since the FET has a very high input resistance. A commercial circuit using a JUGFET is given in Figure 10.14a. The input impedance is about 5 MΩ.

A more usual alternative is to use a configuration of two bipolar transistors known as a *Darlington pair*. The basic connections are shown in Figure 10.14b, and a commercial amplifier (easier to follow than the one in Figure 10.14a!) is given in Figure 10.14c.

This particular circuit is an interesting one. The two transistors of the Darlington pair give a high input impedance, and this stage is capacitor-coupled to the second stage—note that it is unstabilised! A feedback capacitor (20 nF) is used to control the frequency response. The design is simple and is suitable for a wide range of small-signal applications; it was used as a tape reader preamplifier, but could form the basis of an interesting construction project.

fig 10.14

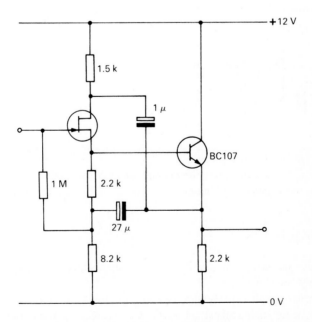

(a) *a commercially designed amplifier using a FET to provide a very high input impedance*

fig 10.14 cont.

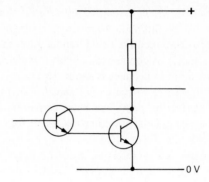

(b) *a Darlington pair of transistors provides high input impedance and considerable gain*

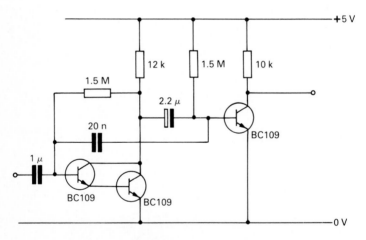

(c) *another commercially designed circuit, using a Darlington pair at the input*

10.7 OSCILLATORS

It is often necessary to generate a continuously changing voltage or current. The frequency of change depends on the applications, but could be anything from one cycle in several minutes to hundreds of megahertz. A circuit that generates such a signal is known as an *oscillator*. About the simplest is the relaxation oscillator, demonstrated in the circuit in Figure 10.15a. This rather old-fashioned circuit makes a good demonstration because you can *see* it working! It uses a *neon lamp*. The neon lamp is filled with neon gas at low pressure. When sufficiently high voltage is applied to the lamp terminals, the neon gas begins to conduct electricity, and at the same time glows red. The lamp will continue to conduct (and glow) until the voltage drops to a level rather lower than that required to 'strike' the neon. It is important to notice that the voltage required to initiate conduction is higher than the voltage required to keep the lamp on.

When the circuit is connected to a voltage source as shown (a high voltage battery, for example) the capacitor C is uncharged. It slowly charges up via resistor R, and the voltage across it increases. At the point where the voltage across the capacitor (and, of course, across the neon lamp terminals) reaches the *striking voltage* of the neon, the lamp abruptly turns on.

fig 10.15

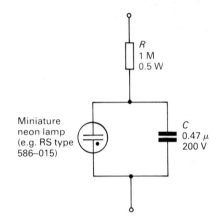

(a) *a simple oscillator, using a neon lamp and capacitor*

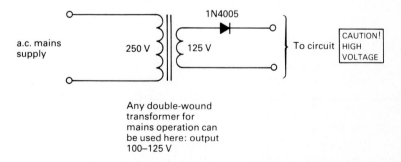

(b) *a suitable power supply for the circuit in Figure 10.15a;*
the diode must have a working voltage of at least 200 V

Current from the capacitor now flows through the lamp, which rapidly discharges the capacitor until the voltage across it drops below the level required to sustain the lamp. The lamp goes out and the cycle repeats.

The circuit shown in Figure 10.15a is useful for demonstrations, but requires a d.c. power source of about 150 V. A suitable power supply is shown in Figure 10.15b. For safety, this circuit should not be used direct from a.c. mains. This illustrates the main reason that the neon oscillator is little used—modern circuits tend to be low voltage. Also, the frequency of oscillation is subject to all sorts of variables; supply voltage, temperature, and even the amount of light falling on the neon lamp will change the frequency.

10.8 THE BISTABLE MULTIVIBRATOR

A circuit which can assume two stable states is known as a *bistable* circuit. Such a circuit is illustrated in Figure 10.16. Essentially, it consists of two simple transistor amplifier circuits, connected so that each transistor's base is connected, through a resistor, to the collector of the other.

fig **10.16** *a bistable multivibrator circuit*

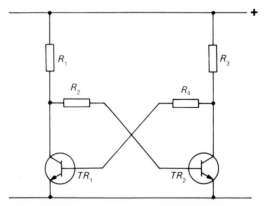

If TR_1 is conducting, its collector will be only a 0.7 V or so above 0 V, and TR_2 is therefore switched off; it is in the non-conducting state. TR_2, in contrast, has its collector at a voltage approaching that of the supply. TR_1 base is therefore held at a high potential which keeps it on. The circuit is clearly *stable* in this configuration, and remains with TR_1 on and TR_2 off indefinitely.

The circuit is, however, symmetrical, and can equally well be stable with TR_1 off and TR_2 on.

Bistable circuits like this form the basis of many sorts of digital memory elements, as we shall see in Chapter 21.

10.9 THE ASTABLE MULTIVIBRATOR

Next, look at Figure 10.17. Assume TR_1 is on; it can only hold the base of TR_2 at a low potential until C_1 is fully charged. R_{B_2} now turns TR_2 on, and the voltage applied to the base of TR_1 drops. It helps to think of R_{B_1}, C_2, and the collector-emitter junction of TR_2, and to consider the changes in potential of the various points in the circuit when the capacitor is charged and uncharged, and TR_2 on or off. See Figure 10.18. TR_1 turns off and the circuit assumes the second 'stable' state. But only until C_2 charges up. Then the circuit swaps over again, TR_1 coming on the TR_2 turning off. This circuit will continue to switch back and forth between the two states at a rate controlled by the circuit values.

fig 10.17 *an astable multivibrator*

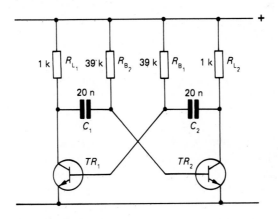

fig 10.18 *one 'leg' of the astable multivibrator*

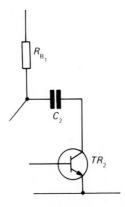

The low frequency limit is generally the size and leakage current of the capacitor, and the high frequency is limited by the type of transistor used. The circuit shown oscillates at around 1 kHz—an oscilloscope, high-impedance earphone, or even a small speaker in series with a resistor, will soon show it working.

Figure 10.19 compares the waveform of the neon oscillator and the astable multivibrator circuits. There are many other types of oscillator circuit in use, though the astable multivibrator is the most common. In digital circuits it is by far the most useful type, as its squarewave output is suitable for driving logic and counting circuits.

10.10 *LC* OSCILLATORS

Figure 10.20 illustrates a simple circuit consisting of a capacitor (*C*) and an inductor (*L*). If a direct voltage power supply is connected to the circuit as shown, the capacitor will charge up, but little current will flow through the inductor (see Chapter 3), at least at the instant of connecting the supply. If the power supply is immediately disconnected, energy will be stored in the capacitor.

The capacitor now discharges into the inductor, and the energy stored in the capacitor is converted into the magnetic field associated with the inductor. After a time, the capacitor will be discharged, and the current flow will stop. But with nothing to sustain it, the magnetic field will begin to collapse, converting its energy back into electricity and recharging the capacitor! While this is happening, current flows round the circuit in the opposite direction.

With the capacitor charged again, the circuit is back in its original state, except that a small amount of energy will have been lost, ultimately radiated by the inductor as a tiny amount of heat. The cycle of charge-discharge repeats continuously until all the energy has been lost. Figure 10.21 gives a graph of current flowing through the circuit.

Notice that the *frequency* of oscillation is constant. Every combination of inductor (*L*) and capacitor (*C*) has its own resonant frequency, in the same way that a clock pendulum has a resonant frequency. Altering either factor (the length of the pendulum or the weight of the bob) alters the rate of oscillation.

The *resonant frequency* of the *LC* oscillator can be calculated as follows:

$$f = \frac{1}{2\pi\sqrt{LC}}$$

(where *f* is the frequency in Hz).

It is possible to use the natural resonance of *LC* circuits to make an oscillator that has a very stable frequency, and usually a sinewave output.

fig 10.19

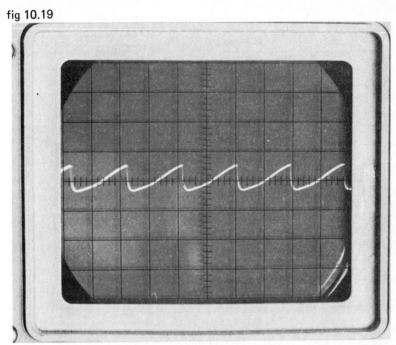

(a) *the waveform produced by the relaxation oscillator in figure 10.15*

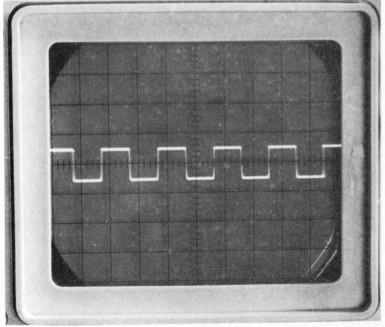

(b) *the waveform produced by the astable multivibrator*

fig 10.20 *a simple LC circuit*

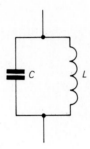

fig 10.21 *voltage across the LC circuit after it is energised; note that although the amplitude of the waveform decays to nothing, the frequency remains constant*

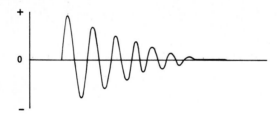

The requirement for oscillation is some form of positive feedback, applied at the resonant frequency by the *LC* circuit. Several designs are routinely used to achieve this. Figure 10.22 shows one of these, the *Hartley oscillator.*

The 100 nF capacitor provides the necessary feedback, while *LC* controls the frequency. This circuit needs an inductor that has a *tapped winding*, i.e. a connection made to the coil in the centre as well as to the ends. Otherwise, it is a simple and reliable circuit. The values shown on the circuit are suitable for an audio oscillator—*L* and *C* should be chosen to give a frequency of 1–10 kHz, if you wish to make up a demonstration circuit.

Another *LC* oscillator circuit is the *Colpitts oscillator.* This does not require a tapped inductor. A typical circuit—again, with values for audio frequency operation—is shown in Figure 10.23. The equivalent of the centre tap is provided by the two capacitors, C_1 and C_2, which (in series combination) give the '*C*' part of *LC*. The coupling capacitor, 10 μF in this circuit, should be large in relation to C_1 and C_2, so as not to influence the frequency.

fig **10.22** *a Hartley oscillator*

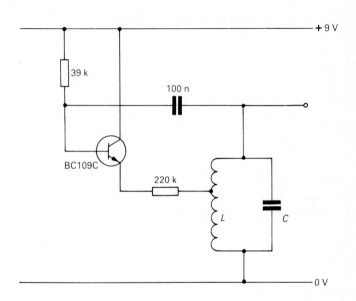

fig **10.23** *a Colpitts oscillator, used in place of the Hartley oscillator where it is not practical to have a tapped inductor*

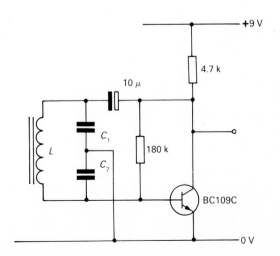

The circuit shown is for operation at audio frequencies; the combined value of C_1 and C_2 should not be more than 100 nF or so.

10.11 CRYSTAL CONTROLLED OSCILLATORS

Although LC oscillators can be made quite stable, the frequency of operation is affected to some extent by temperature and voltage fluctuations. Both the capacitor's capacitance and the inductor's inductance are altered by temperature changes, for example.

The search for a really stable oscillator led to the development of the *quartz crystal oscillator.* Central to this is a specialised component, the quartz crystal. When a crystal of quartz is subjected to an electric voltage, it flexes. Conversely, when you bend the crystal, a small electrical voltage is generated across the crystal. This direct conversion of mechanical to electrical energy (and vice versa) is known as the *piezo-electric* effect. Crystal pick-ups for record players make use of the principle.

Quartz makes a very suitable material for an oscillator crystal because it is elastic—like the pendulum, it takes a long time to stop oscillating once it has started. When 'ringing' at its resonant frequency, the quartz crystal is very like the LC circuit. Energy goes in, which flexes the crystal; the crystal 'swings' back, and electrical energy is produced; as it 'swings' back the other way, the voltage is reversed, and a graph of its output voltage would look just like the one in Figure 10.21.

fig 10.24 *for extreme frequency stability, a quartz crystal controlled oscillator is used*

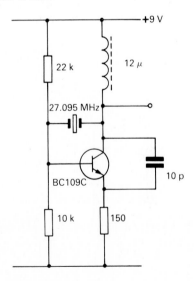

A typical crystal oscillator is illustrated in Figure 10.24. This one works at *radio frequencies*, in the h.f. band—see Chapter 15. The crystal itself provides the necessary feedback between the collector and base of the transistor. Note the use of the inductor in the collector line. This has a low d.c. resistance for the correct biasing levels, but has a high resistance at radio frequencies and allows the output to be taken between the collector and positive supply line. The circuit can be made to oscillate at a frequency determined only by the crystal, regardless of the precise values of other circuit components.

Quartz crystals can be made cheaply and accurately, and in suitable circuits can give astoundingly accurate control. This type of oscillator is used in digital watches, and an accuracy of ten seconds a month (within the reach of the cheapest digital watch) implies a long-term oscillator stability of better than 1 part in a $\frac{1}{4}$ million. Quartz crystals can be made to work over a range of frequencies from tens of kHz to a few MHz.

10.12 POWER SUPPLY REGULATORS

In Chapter 7, the Zener diode was mentioned as a component that is useful for providing a reference voltage. Many circuits require a very stable supply voltage, and the Zener diode is generally used in this context. The simplest possible circuit is given in Figure 10.25.

fig 10.25 *a simple Zener diode voltage regulator*

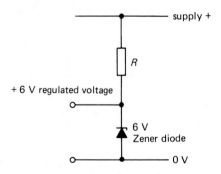

The disadvantage of this is the fact that all the current for the load is supplied through resistor R. The value of this resistor is determined by the power dissipation of the Zener diode, since with the load disconnected all the current flowing through the resistor also flows through the diode. This circuit is therefore limited to applications where very small currents are involved.

A more useful circuit is given in Figure 10.26.

This uses a bipolar transistor to carry the load current, and also to amplify the regulating effect of the Zener diode to provide more accurate regulation. Operation of the circuit is as follows. If there is no load, no current flows through the collector–emitter junction of the transistor. The only current is that flowing through the resistor and the diode, just a few milliamps.

fig 10.26 *a voltage regulator providing improved regulation and increased output current compared with the circuit in Figure 10.25*

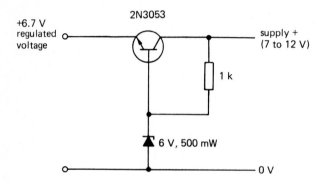

When a load is connected, current flows through the transistor, since the base is held at (in this case) 6 volts, which biases the transistor into conduction. All the while the voltage at the emitter is at or below 6.7 volts (allowing for the voltage drop across the transistor junction) the transistor remains forward biased. But the voltage at the emitter can never rise above 6 volts, since this would involve a higher voltage on the emitter than on the base, reverse-biasing the junction and causing the transistor to switch off. In practice, the voltage on the emitter rises to about 6.7 volts and stays there. If the input voltage to the regulator changes, this does not affect the regulated voltage unless, of course, the input voltage drops below 6.7 volts.

Similarly, changes in the resistance of the load will result only in a compensating change in the bias conditions of the transistor, the voltage at the emitter remaining substantially constant. This type of regulator is useful and inexpensive. The output current is limited by the current and power rating of the transistor.

It is convenient to fit the whole of the regulator circuitry on an integrated circuit, and such components are available cheaply. Figure 10.27 shows a 5-volt regulator circuit, capable of handling currents up to 1 amp.

The performance of the IC is considerably superior to that of the circuit in Figure 10.26, with more accurate regulation, short-circuit protection and automatic shut-down in the event of the IC overheating.

fig 10.27 *a high performance 5-volt regulator using an integrated circuit*

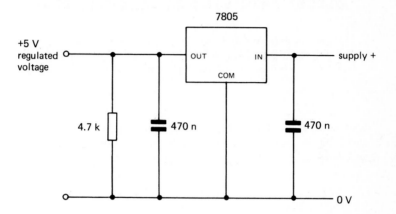

Voltage regulator designs are many and varied, but these days most are based on ICs. High-performance versions use switch-mode regulation, in which the voltage source is switched on and off at a high frequency, the ratio of 'on' to 'off' (the mark–space ratio) changing according to the demands of the load. Such regulators are much more complex, but are more efficient in that little waste heat is produced.

QUESTIONS

1. Why is a bias resistor (or network) required for even the simplest audio amplifier using a single transistor?

2. Why is a capacitor commonly used to couple stages of a multistage transistor amplifier?

3. If an amplifier has a voltage gain of 20 per stage, and it has three stages, what output signal will result from an input of $150\,\mu V$?

4. Which types of oscillator give the best frequency stability?

5. What kind of output waveform is obtained from (i) an astable multi-vibrator, (ii) a relaxation oscillator, (iii) an LC oscillator?

CHAPTER 11

FABRICATION TECHNIQUES AND AN INTRODUCTION TO MICROELECTRONICS

Chapter 7 covered the basic physics of semiconductor devices, and just touched on the methods used to manufacture the components themselves. Semiconductor manufacture, or 'fabrication' as it is more generally called in the industry, is a highly specialised and very difficult subject, but it is useful for the electronics engineer or technician to have some idea of the principles involved.

Various different semiconductor materials are used, but the most common (and cheapest) is silicon. In this chapter we shall look at the fabrication of semiconductor devices based on silicon, but you should bear in mind that roughly similar techniques—although with different materials—are used to deal with germanium and the other semiconductors.

11.1 PURE AND VERY PURE . . .

The first step is to take a single large crystal of pure silicon. This is not as easy as it seems, for silicon that only thirty years ago would have been called 'chemically pure' would be hopelessly contaminated for the purpose of semiconductor manufacture. In the early days of transistor manufacture (1960s) the purity of the raw material, usually germanium, was the major obstacle to reliable production. A process that had worked perfectly for weeks would suddenly start turning out 100 per cent rejects, and would have to be stopped. The batch of semiconductors would be thrown out, everything cleaned, and (with luck) the process might be restored to correct operation after a month. This sort of thing was one of the main reasons for the high cost of early transistors. Today, manufacturers of semiconductor-grade silicon aim for *no* impurity atoms at all. This level is never reached, but crystals having impurities less than 1 part in 1 000 000 000 are routinely made.

One way of making pure silicon crystals was described at the end of Chapter 6. We begin with the sausage-shaped single crystal of silicon that is drawn out of the pure silicon 'melt'.

The silicon crystal is sliced up (like salami) into circular *wafers*, typically 50 mm diameter and 0.5 mm thick. The surfaces are ground and polished perfectly flat, leaving the wafer about 0.2 mm thick. The wafer is finally cleaned with chemical cleaners.

The technique for making a single *pn* diode is as follows.

Beginning with a wafer of *p*-type silicon, made by adding a tiny amount of *p*-type impurity such as indium or boron to the pure silicon, an *n*-type *epitaxial layer* about 15 µm thick is 'grown' on the wafer. This is done by heating the wafer to about 1200°C in an atmosphere of silicon and hydrogen tetrachloride with a trace of antimony, phosphorus or arsenic. Next a thin (about 0.5 µm) layer of *silicon dioxide* is grown on top of the wafer by heating it to about 1000°C in an atmosphere of oxygen or steam. Silicon dioxide has three very useful properties. It is chemically rather inert and is not attacked by gases in the atmosphere, or indeed most other chemicals, even at high temperatures. Second, it is impervious and prevents the diffusion of impurities through it. Third, it is an excellent electrical insulator. The wafer at this stage is shown in cross-section in Figure 11.1.

fig 11.1 *silicon wafer in cross-section, after formation of a silicon dioxide layer on its surface*

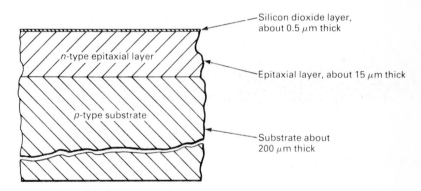

Next, openings have to be made in the silicon dioxide layer, in the right places. This is done photographically. The wafer is coated with a *photoresist*, a light-sensitive emulsion similar to the emulsion on a black-and-white film. A pattern, having the necessary cutouts, is placed over the photoresist, and the wafer is exposed to powerful ultraviolet light. The pattern, or *mask*, is removed, and the wafer washed with *trichlorethylene*, a chemical that dissolves the photoresist only in the places where it was *not* exposed to the ultraviolet light. This leaves the photoresist as a 'negative' of the mask.

At the end of this stage, the wafer is washed in *hydrofluoric acid*, a very powerful acid that will actually dissolve glass—but not the special photoresist (nor, fortunately, the silicon wafer!). The wafer is washed again, and the resist removed with hot *sulphuric acid*. The various steps are shown in Figure 11.2.

fig **11.2** *the various stages in masking and etching the silicon dioxide layer*

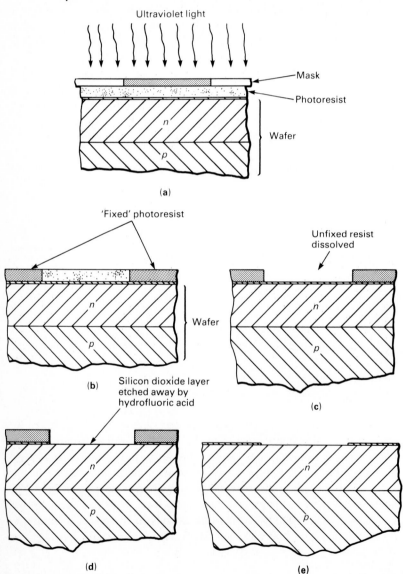

The end-result of this procedure is to produce the right pattern of holes in the silicon dioxide layer. Our wafer now goes back in the furnace, with an atmosphere of *p*-type impurity. The impurity diffuses into the wafer below the holes in the silicon dioxide, but nowhere else (see Figure 11.3).

fig 11.3   *p-type impurity is diffused into the upper n-type layer*

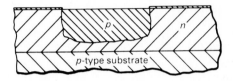

The whole process is repeated, with a different mask, to diffuse another *n*-region into the *p*-region just created. Then the whole process is repeated a *third* and then a *fourth* time, to diffuse more doped regions into the wafer. Figure 11.4 shows the final result.

fig 11.4   *a completed silicon planar epitaxial pn diode*

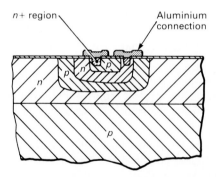

Connections have been made to the heavily doped *n*+regions, by evaporating an aluminium film on to the wafer, after suitable masking. If this seems a terribly complicated way of making a *pn* diode (compare Figure 7.7, p. 84, with Figure 11.4, for example) then it is for a reason.

Those extra *p* and *n* layers below the diode itself—the *pn* junction connected to the outside world—serve to isolate the diode from the rest of the wafer. A reverse-biased *pn* junction surrounds all parts of the diode, preventing leakage to the rest of the wafer. Now this is not a very useful feature if we are going to cut the wafer up into individual diode *chips*. But suppose we cut the wafer into larger sections? The aluminium layer could carry current from one part of the wafer to another, and transistors and diodes could all be combined on a single chip to build a complete circuit!

And this idea is the basis of microelectronics. The first integrated circuits used just this technique, though more recent devices are rather more subtle, and rely on electrical connections inside the silicon structure rather than on superimposed aluminium 'wiring'.

11.2 INTEGRATED CIRCUITS

Diodes and transistors—bipolar and FETs—can be produced on a silicon wafer. So, too, can resistors, either deposited on top of the wafer in the form of *tantalum* (a poor conductor), or built into the wafer as a 'pinch' resistor. The pinch effect is similar to that observed in a partially turned-off FET (see Figure 9.6, p. 113) and relies on a very thin region for conduction. Resistance values of up to $100 \text{ k}\Omega$ can be produced in this way.

Capacitors are more problematic, and although it is possible to make capacitors on an integrated circuit, the values are generally limited to a few picofarads if the capacitor region is not to be excessively large. Capacitors are to be avoided in integrated circuits as far as possible.

There is no equivalent of an inductor, but fortunately most circuits can be designed to avoid this requirement.

11.3 TESTING AND PACKAGING

The silicon wafer may eventually contain several hundred complete integrated circuits. They are, of course, all produced simultaneously, the masks used during processing consisting of hundreds of identical units. See Figure 11.5.

It is usual to test the individual circuits before the wafer is cut up. Testing is done automatically, and the machine marks any faulty circuits with a blob of ink. The *yield*, i.e. the percentage of 'good' circuits, varies considerably. Complex circuits might have a yield as low as only a few per cent, whereas simple circuits would be much better. The reason is clear—if a number of faults are scattered over a wafer, the larger the number of circuits on that wafer, the more fault-free circuits there will be. Once the wafer has been tested, it is cut up and the faulty circuits discarded, again automatically. The chip is mounted on a suitable frame, and connections are made between the aluminium 'pads' on the chip and the frame which forms the connections to the chip. The chip is still usually hand-wired to the frame, using ultrasonic welding to make the connections. The work is done under a binocular microscope, using special micro-manipulators. Fine gold wire is often used for the connections. Figure 11.6 illustrates the first part of the process.

fig 11.5 *a silicon wafer—with a ruler to indicate the scale—showing a little under 200 individual integrated circuits, ready to be tested and cut up*

photograph by courtesy of Ferranti Electronics Ltd

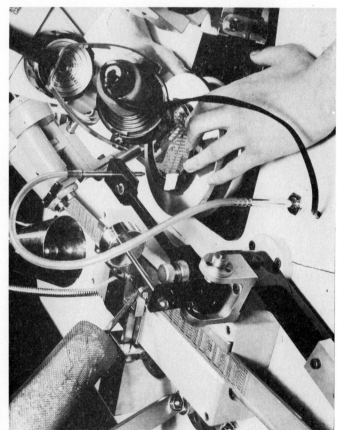

fig 11.6 mounting individual silicon chips in their frames, prior to adding the connections and encapsulation

photograph by courtesy of Ferranti Electronics Ltd

Various styles of encapsulation are used, but by far the most common is the DIL pack (Dual-In-Line). DIL packs are based on a standard $\frac{1}{10}$ inch matrix. The pins down each side are always 0.1 inch apart, and the distance between the two rows of pins is a multiple of 0.1 inch, generally 0.3 inches for packages of eighteen pins or less, and 0.6 inches for twenty pins or more. Figure 11.7 illustrates typical plastic DIL packs.

fig 11.7   *a selection of DIL-packs, the form in which integrated circuits are generally sold*

Ceramic DIL packs are used as well as the plastic ones, but only for the most expensive devices, or those requiring a very high degree of environmental protection. The number of individual semiconductor devices that can be packed on to a single chip is astonishing. Although a 14-pin DIL pack may contain a relatively simple circuit (see below) a 40-pin DIL pack can contain the electronics for a complete computer! Figure 11.8 shows a complex integrated circuit. It is in fact the central processor unit of a computer. Devices like this have a huge number of transistors and diodes, approaching 10 000 on a chip less than 10 mm square.

fig 11.8 *an example of a large-scale integrated circuit, a microprocessor;
this IC contains all the important parts of a computer central
processor unit*

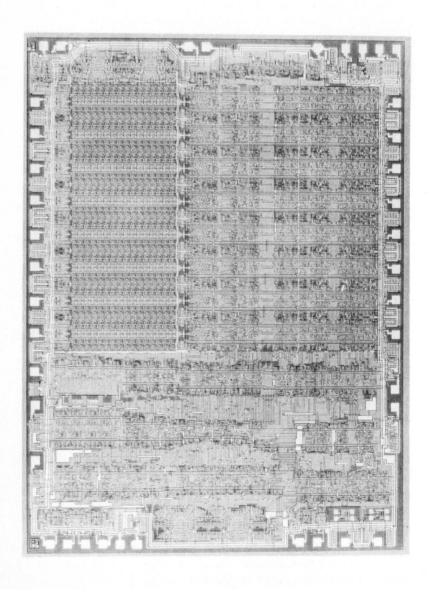

photograph by courtesy of RCA

fig 11.9 *circuit diagram of a typical bipolar IC, operational amplifier type SN72741*

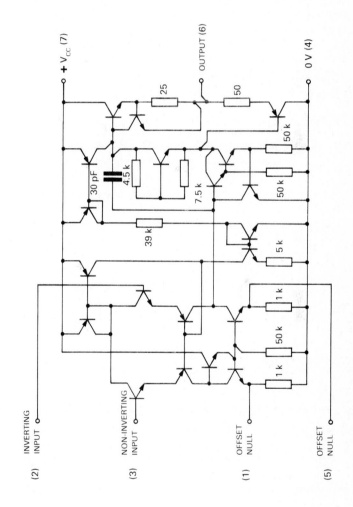

11.4 SOME EXAMPLES OF SIMPLE INTEGRATED CIRCUITS

In the next chaper we shall look at *operational amplifiers*, so it seems sensible to take one of the most popular operational amplifiers as an example of the kind of circuit design used (see Figure 11.9). This is the operational amplifier type SN72741. As you can see, there are many transistors, a few resistors (they take up more room on the chip than transistors or diodes), and as few capacitors as possible (one). But this is a relatively simple circuit; it is very cheap, currently costing less than a cup of coffee. Such is the scale of mass production in the microelectronics industry!

The second example—in Figure 11.10—relates to Chapter 19, and shows a 2-input AND gate. A 14-pin DIL pack contains a chip with four such gates.

fig 11.10   *a typical simple IC using MOSFET technology; components can be packed up to twenty times more densely than with bipolar technology (this diagram shows one of four identical circuits included in the CD4082 quad 2-input AND gate)*

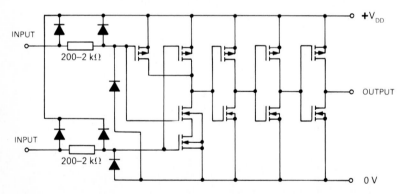

The use of MOSFET technology means greater economy, as it can be packed on the chip up to twenty times more densely than bipolar transistors. Also, the power requirements are much lower. The CD4082 quad 2-input AND gate is as cheap as the SN72741 operational amplifier.

More complicated chips, such as the Z80 microprocessor, are more expensive, but still amazingly cheap. The Z80 currently costs about the same as a pound of coffee beans.

QUESTIONS

1. Why would a complex IC have a lower production 'yield' than a simple one?

2. What method is used to manufacture pure silicon such as might be used for semiconductor manufacture?

3. Silicon dioxide layers feature prominently in the manufacture of all silicon semiconductor devices. What are important properties of silicon dioxide?

4. What is the pin spacing of the standard DIL pack?

5. Name two 'impurity' materials that are added to pure silicon to produce (i) n-type, (ii) p-type silicon.

OPERATIONAL AMPLIFIERS

Although the transistor amplifiers we looked at in Chapter 10 are useful for all sorts of applications, designers of amplifier systems are turning more and more to integrated circuit designs. The *operational amplifier* (the name goes back to the days when *analog* computers were more widely used, but that's another story . . .) is perhaps the basic building-block of linear electronic systems. The 'op-amp' (a commonly used abbreviation) is designed to be a close approximation to a perfect amplifier. Here is a specification for such a perfect device:

1. **Gain**. This should be infinitely high. Obviously an amplifier with infinite gain would be useless, as the smallest input would result in full output! As we shall see, a very high or even infinite gain can be controlled by suitable feedback.
2. **Input resistance**. Ideally, this should also be infinite, so that there is no loading of the input source at all.
3. **Output resistance**. Ideally, this should be zero. With a zero output resistance, the amplifier can be connected to a load of any resistance without the output voltage being affected.
4. **Bandwidth**. This should be infinite, which means that the amplifier should be able to amplify (infinitely!) any frequency from zero (direct voltage) to light!
5. **Common mode rejection ratio**. This, too, should be infinite, but an explanation is needed. An operational amplifier has one output, but two inputs, an *inverting input* and a *non-inverting input*. Figure 12.1 shows an op-amp system, with the two inputs clearly marked, '−' for inverting and '+' for non-inverting. A positive voltage applied to the inverting input makes the output swing negative and a positive voltage applied to the *non*-inverting input makes the output swing positive. It is vital to note that the inputs are relative to *each other* and not to either of the supply lines. Thus, if *both* inputs (common mode) are

fig **12.1** *circuit symbol for an operational amplifier*

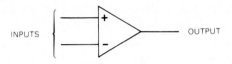

fig **12.2** *showing the output of the op-amp for various types of input*

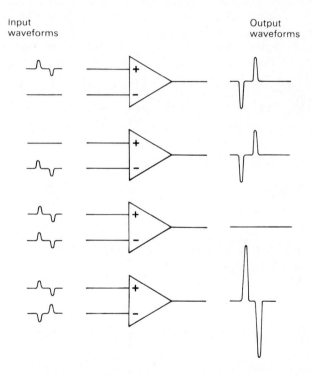

made more positive or negative, there should be no output. This is illustrated in Figure 12.2.

6. **Supply voltage**. The amplifier should be unaffected by reasonable variations in its power supply voltage.

Now we can compare the theoretical specifications with a real one–the specification for the SN72741 op-amp:

1. **Gain**: 200 000 voltage gain, about 106 dB.
2. **Output resistance**: 75 Ω.
3. **Input resistance**: 2 MΩ.

4. **Bandwidth**: d.c. to 1 MHz.

5. **Common mode rejection ratio**: 90 dB (i.e. a signal applied to both inputs will be at least 32 000 times smaller than the same signal applied to one input).

6. **Supply voltage**: output will change less than $150\,\mu V$ per volt change in the power supply. The amplifier will operate from a supply of + and −3 to 18 V (see below) and takes a quiescent supply current of about 2 mA. The maximum power dissipation is 50 mW, and the maximum output current is around 30 mA, by no means perfect, but not bad for a device retailing at less than the price of a cup of coffee! The SN72741 is far from being a 'state of the art' device, but it is a good general-purpose op-amp, ideal for all sorts of commercial (i.e. cost-effective) designs, and very suitable for demonstrations and experiments.

Figure 12.3 shows the connections to this integrated circuit, which is most commonly available in an 8-pin DIL pack.

fig **12.3** *the SN72741 op-amp integrated circuit*

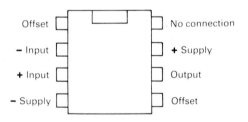

The pins marked 'offset' can for the most part be ignored. They are there to compensate for a small design problem with the SN72741, namely the fact that with both inputs at exactly the same potential the output may be a fraction positive or negative of zero. Usually this does not matter, but in applications where it does the simple arrangement shown in Figure 12.4 provides a pre-set control for exact setting.

Despite the usefulness and wide application of the op-amp, there will be many occasions when the engineer will want to design an amplifier using discrete components. There will be even more occasions when the service engineer will meet other types of amplifier. But the study of operational amplifiers is important because the general principles of design—particularly in feedback circuits—applies to virtually *every* amplifier design.

fig **12.4** *offset null setting for the SN72741*

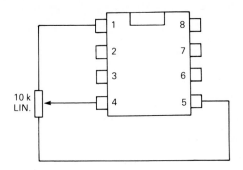

12.1 NEGATIVE FEEDBACK TECHNIQUES

First of all, we need some control over that (almost) infinite gain! Figure 12.5 illustrates the basic op-amp configuration with feedback.

fig **12.5** *the basic op-amp feedback configuration, using the inverting input*

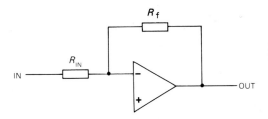

Assuming that the amplifier really does have infinite gain, the gain is controlled only by the values of the resistors R_{IN} and R_f. R_{IN} is the input resistor, and must be substantially less than the input resistance of the op-amp. R_f is the feedback resistor. The voltage gain of the system is very simply calculated as:

$$A = \frac{R_f}{R_{IN}}$$

where A is the voltage gain. The fact that the op-amp has finite gain affects the calculation only slightly, provided the required gain is not approaching the specified maximum of 100 dB or so. Figure 12.6*a* is a practical circuit. Notice the odd power-supply requirements. The SN72741 needs a power supply that is symmetrical about zero. This permits the output voltage to swing above and below zero. There are various ways of contriving such a supply, but the simplest (and good enough for our purposes) is to use two 9 V batteries (PP3 or PP9) wired as shown in Figure 12.6*b*.

fig 12.6

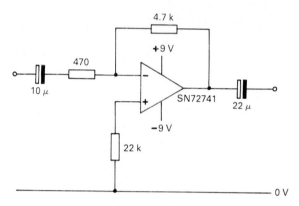

(a) *a practical amplifier based on the circuit of Figure 12.5*

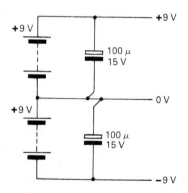

(b) *a simple power supply that can be used for the amplifier in Figure 12.6a*

Remember that the inverting input is relative to the non-inverting input, not to the 0 V supply line. In our simple amplifier we want the input to be relative to 0 V ('earth'), so we connect the non-inverting input to 0 V. This is best done via a resistor, though the value is uncritical— 22 kΩ is convenient. This amplifier circuit has a gain of ten time (3 dB) set by the values of the input and feedback resistors, 4700/470. The capacitors are added for a.c. operation, and could be left out for low frequency applications, according to the input characteristics.

This amplifier is inverting. The design of a non-inverting amplifier is slightly more difficult, since although the input is applied to the non-inverting input, the feedback still has to be applied to the inverting input. The basic configuration is shown in Figure 12.7.

fig 12.7   *the basic non-inverting configuration for the op-amp*

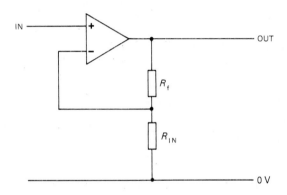

The gain of the non-inverting amplifier shown here is calculated as:

$$A = \frac{R_{IN} + R_f}{R_{IN}}$$

A practical non-inverting amplifier with a gain of 11 is shown in Figure 12.8. The capacitors are added for a.c. operation, as for the circuit in Figure 12.6.

It is possible to apply several inputs to the op-amp configuration, isolating the inputs from one another with the input resistors, which should be as high in value as practicable—the higher they are, the better the isolation. Such a circuit could be used for an audio mixer, shown in Figure 12.9. The input resistors R_1, R_2 and R_3 set the maximum gain of the channels; in this case, R_1 and R_2 give a gain of 10, and R_3 gives a gain of unity (no amplification). The logarithmic potentiometers give volume controls that are suitable for audio use—slider controls are more convenient to use than rotary ones.

fig 12.8 *a practical circuit based on the configuration in Figure 12.7; the power supply in Figure 12.6b can be used for this circuit*

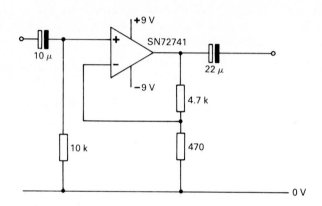

fig 12.9 *a practical circuit for an audio mixer; an op-amp provides amplification (the power-supply circuit of Figure 12.6b can be used with this design)*

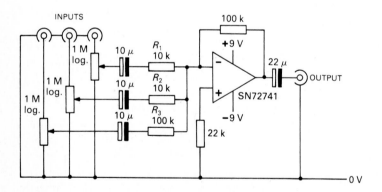

12.2 POSITIVE FEEDBACK TECHNIQUES

It may appear at first that positive feedback—feedback that reinforces the input rather than reduces it—would not have a great deal of application to a very high gain operational amplifier. This is not in fact the case, and a whole class of circuits is based on positive feedback. Figure 12.10 shows the basic configuration for positive feedback.

The similarity to Figure 12.7 is superficial—compare the positions of the inverting and non-inverting inputs. Assume there is no output

fig **12.10** *an op-amp in a positive feedback configuration; note that it is the non-inverting input that is connected to the junction of the two resistors*

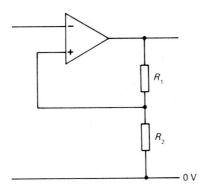

from the circuit, the output being exactly zero. If a small positive voltage is applied to the input, the output swings negative. The non-inverting input, previously held at zero volts through R_2, is now provided with an amplified negative signal from the output, via R_1. This reinforces the input by increasing the potential difference between the two inputs. The result is that the output swings very rapidly to its maximum negative excursion, just slightly away from the negative supply voltage.

If a sufficiently large *negative* potential is applied to the input, enough to make the output positive, even momentarily, the circuit will rapidly change state and the output will swing to its maximum positive excursion. This is very like the bistable circuit of Chapter 10, in that the circuit can adopt one of two stable states: this is, in fact, an op-amp bistable. A practical circuit is shown in Figure 12.11, the only extra factor being the extra resistor to ensure that the inverting input remains at zero volts in the absence of an applied signal.

An adaptation of this circuit can be used as a perfectly usable touch-switch, to turn a lamp on or off. Figure 12.12 provides a suitable circuit.

The touch contacts are illustrated in Figure 12.13 and can be made with drawing-pins. The resistance across a human thumb is about $50\,k\Omega$ to $2\,M\Omega$, depending on the dryness of the skin, and this allows enough current to flow to make the circuit change state. The tiny current is, of course, completely harmless, but the circuit must be used only with battery power or with a correctly designed *and isolated* power supply. The diode is necessary to ensure that the lamp lights only when the output of the amplifier is negative; without the diode, the lamp would light all the time.

fig **12.11** *a practical circuit for an op-amp bistable*

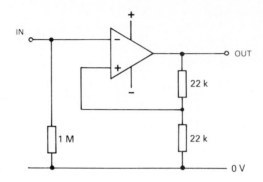

fig **12.12** *a practical design using the circuit of Figure 12.11 that operates as a touch-sensitive switch to turn a lamp on or off; if the circuit is required to switch a high voltage or current, the lamp can be replaced with a relay (see Chapter 17)*

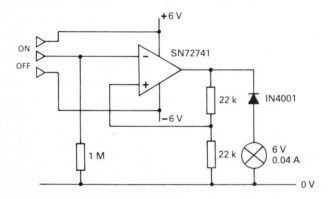

fig **12.13** *suggested layout for the switch for Figure 12.12*

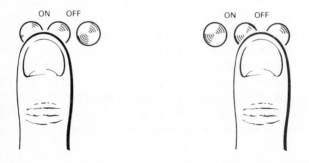

12.3 OPERATIONAL AMPLIFIER OSCILLATORS

In Chapter 10 we developed the bistable multivibrator into an astable. There is a parallel with the op-amp. Figure 12.14 shows a basic op-amp oscillator.

fig 12.14 *an op-amp oscillator*

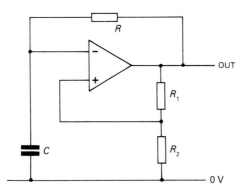

The principle is straightforward. Starting with the amplifier in a condition where the output is positive, the capacitor C is charged, via R, until the inverting input becomes sufficiently positive to cause the 'bistable' to change state, the output now becoming negative. It is now a negative potential that is applied to C, via R, and C is discharged and recharged with the other polarity. When the potential on the inverting input charges, the 'bistable' again changes state; and so on. The calculation for the rate at which the circuit changes state is given by the following formula, where T is the total time taken for the oscillator to go through one complete cycle:

$$T = 2RC \ln \left(1 + \frac{2R_1}{R_2}\right)$$

(ln is the natural logarithm, log to the base e). Figure 12.15 gives a practical circuit to demonstrate the op-amp astable, with a simple transistor amplifier to increase the output volume.

The variable resistor allows you to adjust the output over a range of audio frequencies. The output of the circuit is approximately a square-wave (Figure 12.16).

One type of oscilator that produces a sinewave output is the *Wien bridge oscillator*. This circuit is shown in Figure 12.17. This circuit has

fig 12.15 *a practical circuit enabling the op-amp oscillator to drive a small speaker; the operating frequency of the oscillator can be adjusted by means of a variable resistor*

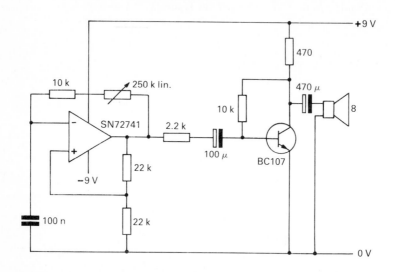

fig 12.16 *the output of the oscillator in Figure 12.15*

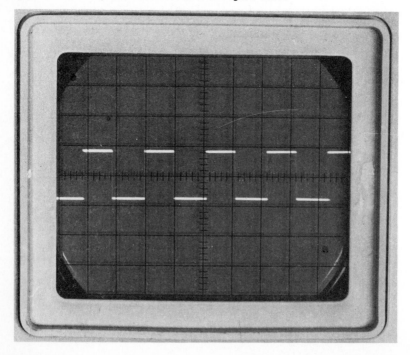

fig 12.17 *the Wien bridge oscillator is used where a sinewave output is required*

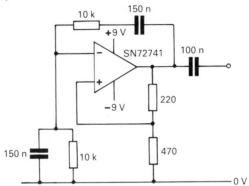

fig 12.18 *the output of the Wien bridge oscillator*

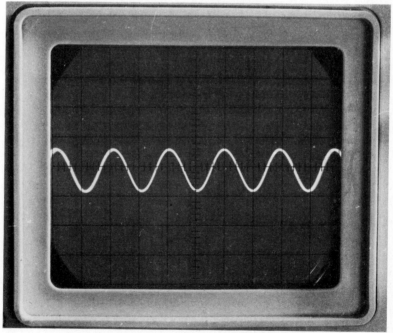

two Rs and two Cs, which makes frequency adjustment difficult. A dual $10\,k\Omega$ potentiometer can be used. The frequency of operation is given by:

$$F = \tfrac{1}{2}\pi RC$$

where F is the frequency in hertz. Figure 12.18 gives the output waveform.

12.4 CONTROL OF FREQUENCY RESPONSE

By putting frequency-sensitive components or networks in the feedback loop, the frequency response of the op-amp can be controlled. If the feedback resistor R_f in Figure 12.5 is replaced by a capacitor, then the op-amp will act as a *low-pass filter*. As high frequencies pass through the capacitor more readily than low frequencies, the feedback is greater at high frequencies and the gain lower.

A resistor and capacitor in parallel, used in the feedback loop—Figure 12.19—will cause the amplifier to amplify a selected frequency, with the gain falling off above and below the centre frequency. The same formula that we used for the Wien bridge oscillator $(F = \frac{1}{2}\pi RC)$ gives the centre frequency, the frequency at which the amplifier exhibits the highest gain.

fig 12.19 *the frequency response of the op-amp can be controlled by means of frequency-sensitive components in the feedback loop*

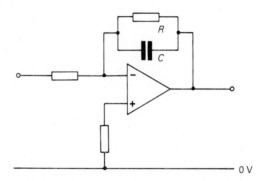

QUESTIONS

1. Give the ideal specification of an operational amplifier in terms of (i) gain, (ii) input resistance, (iii) output resistance, (iv) bandwidth.

2. Draw circuit diagrams for an op-amp using (i) negative feedback, (ii) positive feedback, providing a bistable circuit.

3. Draw circuit diagrams for (i) inverting, (ii) non-inverting amplifiers using op-amps. In both cases show resistor values to give a gain of 25.

4. Draw a circuit (omit component values) for an op-amp astable circuit.

5. What is meant by 'common mode rejection ratio'?

AUDIO AMPLIFIERS

A record-player is one of the simplest electronic systems. The system diagram, in its basic form, has only three components—the pick-up, the amplifier, and the speaker. The system has been illustrated in Figure 10.1 (p. 119).

But audio frequency amplifiers, i.e. amplifiers designed to amplify signals at around the frequencies to which the human ear is sensitive, have their own special problems, and the circuits are designed to solve these problems. The range of frequencies to be amplified is well within the capability of even simple circuits. The generally accepted 'hi-fi' frequency range is between 20 Hz and 20 kHz, though only young children can hear frequencies as high as this, and at frequencies as low as 20 Hz the sound is more *felt* than heard. The first requirement for an audio amplifier to be considered here is the *power output*. A typical transistor radio or portable tape-player will provide an output power of around 500 mW into a small speaker. This sort of power level proves quite adequate for general listening at fairly close range. But for hi-fi, where the faithful reproduction of high-energy transient sounds is important, 20 W is considered a sensible minimum.

Let us return to the portable player. To deliver about 500 mW into the speaker requires about 55 mA from a 9 V supply, if we neglect any inefficiencies in the system. This power output will be required only rarely— the loudest sounds, with the volume turned well up. This appears to be well within the capabilities of a handful of small dry batteries. A very simple way of driving a speaker is shown in Figure 13.1. (The speaker itself is dealt with in detail on pp. 264).

You will (I hope) recall from Chapter 10 that in order to amplify audio signals in an undistorted way, the output of an amplifier has to be able to swing positive and negative of a neutral point.

This is achieved in amplifiers like the one in Figure 13.1 by biasing the transistor so that the collector is at approximately half the supply voltage.

fig **13.1** *a simple way of driving a speaker with a transistor amplifier*

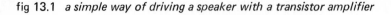

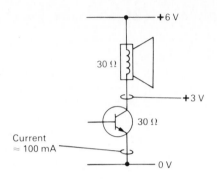

In the simple amplifier in Figure 13.1, this happens when the resistance across the transistor's emitter–collector junction is equal to the resistance of the collector load; the transistor and the load resistor are then the two halves of a potential divider with equal resistance either side, and the mid-point will be at half the supply voltage.

The speaker's coil has a resistance of 30 Ω, so the total resistance is about 60 Ω. These figures are quite realistic. Thus the current taken by this output stage is roughly 6 × 1000/60 = 100 mA. For small and medium-sized dry batteries, this represents a fairly short life. And notice that the current is consumed *even when there is no output* from the speaker. In fact, the current taken from the supply will, on average, always be the same, since for the most part audio signals will reduce and then increase the transistor's emitter–collector resistance symmetrically.

There is also the question of what happens to the power lost from the batteries. Half is dissipated by the transistor, and the other half by the speaker—300 mW each. Thus cooling, even in a modest amplifier, is something that needs to be taken into consideration.

And finally there is that 100 mA flowing continuously through the speaker, pulling the speaker cone out of its true central position. It means that a small speaker could not be used, and for this (and other) considerations, the whole amplifier is going to be quite large.

In case this design seems so hopelessly unsuitable, a complete amplifier based on the simple circuit above is shown in Figure 13.2. This was produced commercially in the 1960s, and actually makes a good, musical sound. It would be quite an interesting project to make it, as the transistors (or their equivalents—the circuit is pretty tolerant) are still available.

The transistor will need a heat sink of some sort (a piece of aluminium 1 mm thick by 100 mm square would do) and a speaker of at least 150 mm diameter is recommended. Speakers with 30 Ω coils could be hard to find, so the circuit values have been changed for a 15 Ω speaker. The circuit

fig **13.2** *a practical direct-coupled class A audio amplifier*

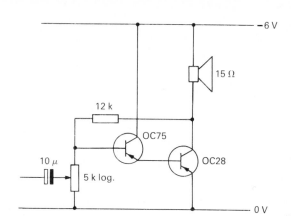

takes about 300 mA from a 6 V battery (a 6 V motor-cycle battery would have a reasonable life!). Use a crystal cartridge, they were widespread in the 1960s and have a high output. This amplifier hasn't a great deal of gain.

The system of volume control is simple, the control working as a potential divider to feed a proportion of the signal to the amplifier, according to the setting. The use of a logarithmic-law control is standard, and gives a smooth increase and decrease in volume. If you make this amplifier as a demonstration, try the effect of a linear-law potentiometer.

The amplifier we have been discussing is known as a *class A* audio amplifier. A properly designed class A amplifier can have very low distortion, but will always waste a great deal of power.

13.1 CLASS B AMPLIFIERS

Today, with a few eccentric exceptions, audio amplifiers are *class B* amplifiers. It is possible (but more difficult) to obtain very low distortion, and the class B amplifier consumes power that is much nearer to being proportional to the output power at any instant. In other words, a loud signal will cause the amplifier to draw a heavy current from the supply, whereas when quiescent (no signal) it will take very little current.

Figure 13.3 illustrates one of the basic forms of the class B amplifier. This is known as a *complementary-symmetry* class B amplifier because it is a symmetrical circuit, and because it uses complementary transistors, a *pnp* and an *npn* type, as the output devices. The circuit depends for its operation on the fact that the transistors require different polarity of signal to drive them. The input waveform is shown in the diagram—one

fig 13.3 *a class B audio amplifier in which the output current is shared between two transistors*

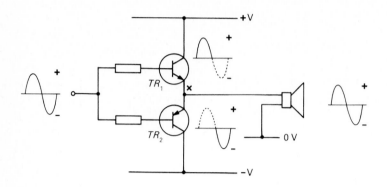

cycle of a sinewave at an audio frequency. The first part of the signal, the positive half-cycle, makes TR_1 conduct and act as an amplifier. TR_2, on the other hand, will be turned off (non-conducting) by the signal. This means that point 'X' on the diagram, the output to the speaker, is the output of the transistor amplifier TR_1. The speaker's speech coil is the emitter load of this amplifier.

The second—negative—half-cycle of the input signal switches TR_1 off, but makes TR_2 conduct. This time TR_2 is the amplifier, with the speaker still the emitter load of this amplifier. The result is that the two halves of the input signal are handled by two different transistors, but the output from both amplifiers is fed to the speaker, to reconstitute the original signal. The efficiency of the amplifier comes from the fact that, with no input signal, *neither* transistor is conducting, so no power is taken from the supply, and no energy needs to be dissipated by any part of the circuit (or the speaker) when there is no sound coming out of the amplifier.

Like most things, it turns out that the real circuit isn't as simple as basic theory seems to suggest. The load lines in Figure 8.12 (p. 103) suggest the nature of the problem: the fact that the transistors do not respond to a signal in a linear manner when operated near cut-off point. The transistors have to be biased so that they are just conducting enough to move them into the linear part of the characteristic. Distortion will then be minimised. In class B amplifiers it is distortion at the point where one transistor takes over from the other that is most often encountered, especially in cheap or poorly adjusted designs. Such distortion is called *crossover distortion*. An example of crossover distortion in a poorly adjusted design is shown in Figure 13.4. The input is a sinewave, in this test.

fig **13.4** *crossover distortion*

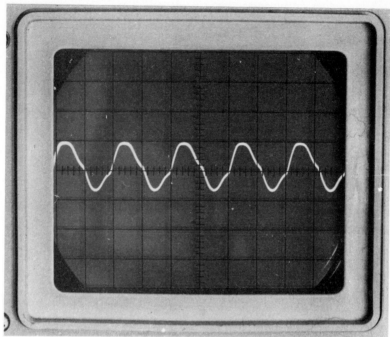

A circuit with a suitable bias system to maintain the transistors at the correct point on the curve is illustrated in Figure 13.5. The diodes have a secondary purpose, that of providing the amplifier with thermal stabilisation. For the lowest possible crossover distortion, the operating points of the transistors need to be set accurately. As the amplifier operates, particularly if it is delivering a large output, the transistors will get hot, changing the characteristic curves and necessitating a slightly different bias setting.

Automatic compensation is obtained by putting the diodes in contact with the transistors, so that the increase in temperature of the transistor also heats the diode. The diode's conduction changes correspondingly, altering the bias conditions so as to compensate (approximately) for the transistor's changes. The system is surprisingly effective. Notice that the amplifier in Figure 13.5 uses a split supply, like the op-amps. This is an efficient mode of working, but the use of a split supply may well be inconvenient in a small portable device. It is possible to use a single supply, but a capacitor must be put in series with the speaker to prevent a direct current flowing all the time. The layout is shown in Figure 13.6. Because of the large currents flowing, the capacitor must be very large, in the order of hundreds of microfarads. Often it will be by far the largest physical component in the circuit.

fig **13.5**   *a practical class B output stage to drive a 15 Ω speaker; the diodes are in contact with the encapsulations of the transistors and provide thermal compensation*

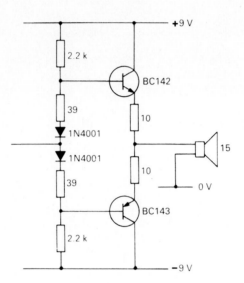

In the design of the class B output stage we have concentrated on power-handling and efficiency. Gain is a low priority, and the amplifier will invariably use more, low-power stages to amplify the incoming signal to a level where it can drive the output transistors. (In the next section of this chapter we will consider *preamplifiers*, as they are called.)

Before leaving the subject of class B power output stages, it is important to realise that the design shown in Figure 13.5 is by no means the only one possible. It is often used because it is simple and works well. However, amplifiers are still in use—paradoxically in cheap radios—that have transformer coupling of the input and output stages, and identical transistors on both sides of the 'push–pull' class B layout. The transformer coupling to the input is used to provide the two halves of the system with input signals of opposite phase. An output transformer may be used to match the speaker impedance to the output stage more closely. A transformer-coupled design is shown in Figure 13.7.

Transformer coupling is now used rarely, since the price of transformers and their size make them undesirable for modern circuits. The transformer also adversely affects the frequency response and distortion, so transformer coupling is not used in hi-fi designs.

fig **13.6** *the requirement for a split power supply can be obviated by the use of an output-coupling capacitor*

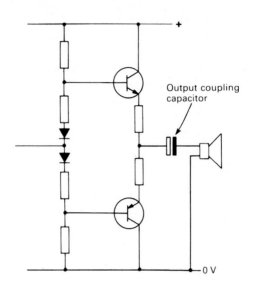

fig **13.7** *a typical transformer-coupled class B amplifier*

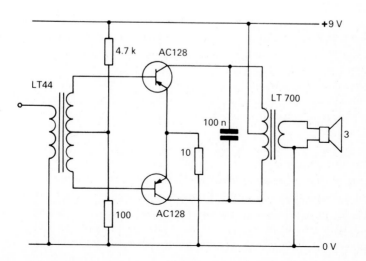

13.2 PREAMPLIFIER AND DRIVER STAGES

Often the class B output stage is preceded by a single transistor amplifier that provides some of the gain needed for the amplifier. This in turn is preceded by a *preamplifier*, which provides most of the gain and may also incorporate tone and volume controls. Any division between the driver and preamplifier is rather artificial, and today the whole system coming before the power amplifier tends to be termed the 'preamplifier'.

A simple preamplifier can be made with the ubiquitous op-amp. Neglecting tone and volume controls for the moment, the design in Figure 13.8 is completely practical.

fig **13.8** *an op-amp preamplifier stage*

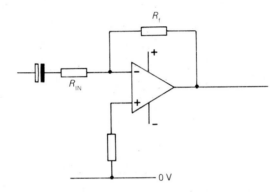

Compare this circuit with the one in Figure 12.6*a* (p. 158); this is a simple inverting amplifier. Capacitor coupling to the power stage is not in fact needed, so the complete preamplifier and power amplifier system is arranged as shown in Figure 13.9. This is in fact an entirely practical circuit, and is an amplifier with a quite respectable performance. The frequency response should be only about 3 dB down at 20 Hz and 25 kHz, it has an input impedance of 10 kΩ, and will deliver about 200 mW into a 15 Ω speaker.

Note that the diodes should be in contact with the cases of the transistors for automatic thermal compensation. The volume-control system is unusual in that it is in the feedback line and works by altering the overall gain of the preamplifier. A more usual system would be to use a potentiometer to proportion the incoming signal, like the one shown in Figure 13.2. However, the feedback system reduces the amplifier's background noise as the volume is lowered, a desirable characteristic. In hi-fi amplifiers

fig **13.9** *a practical audio amplifier, using the op-amp as a preamplifier and a class B output stage; the volume control operates by varying the gain of the preamplifier*

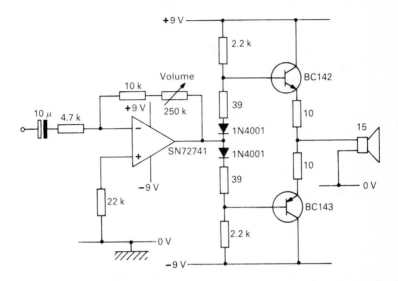

it is usual to design the preamplifier with discrete components. It is possible to build preamplifiers with exceedingly low distortion—verging on the unmeasurable.

13.3 TONE CONTROLS

The very simplest kind of tone control, used in the very cheapest designs, uses a simple 'treble-cut' circuit. A variable resistor in series with a capacitor is placed across the signal line at a convenient point. When the control is set to low resistance, the capacitor, working as a frequency-sensitive resistive component, shorts out a substantial proportion of the high frequency signal. The variable resistor gradually reduces the effect of the capacitor as it is turned up towards maximum resistance. This type of circuit, shown in Figure 13.10, is widely used, though it has little to recommend it apart from cheapness.

The design used in almost every hi-fi amplifier, and most other medium- and high-quality applications, is the *Baxandall* tone-control circuit, named after its inventor. The Baxandall circuit works by selective feedback, and has the advantage that, with comparatively few components, it is possible to control bass and treble separately, and to give cut *and* lift to each as required. A Baxandall tone control circuit using an op-amp is shown in Figure 13.11.

fig **13.10** *a cheap and nasty form of tone control*

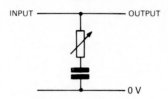

fig **13.11** *a Baxandall tone-control circuit, used in all quality audio equipment; this circuit provides both lift and cut at bass and treble frequencies*

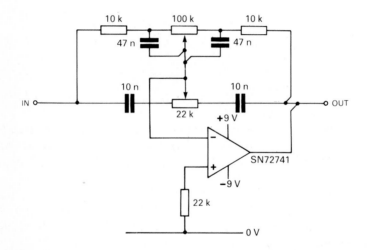

This amplifier has a gain of unity (i.e. with the tone controls 'flat', the output signal is at the same level as the input signal) and, when used in an amplifier system, would be placed between the preamplifier and power amplifier stages.

13.4 INTEGRATED CIRCUIT AMPLIFIERS

The trend in electronics today is towards fewer and fewer components in a given system. Figure 13.12 shows a typical 200 mW amplifier of the late 1960s, with twenty individual components. The circuit is straightforward and quite comprehensible, and you should have no problem in following it. Compare it with Figure 13.9, which has fifteen components and a substantially better performance. The main saving is by using the op-amp.

fig **13.12** *a typical low-power amplifier of the late 1960s; it involves twenty individual components and about forty soldered joints*

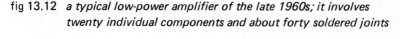

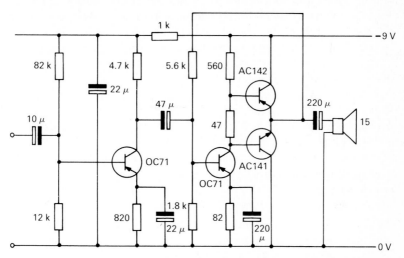

An audio amplifier is just the sort of device that is suitable for production as an integrated circuit—there is a large market for identical units, and all components in the system (most of them, anyway) can be produced using microelectronic techniques. It is not surprising, therefore, that there is no justification nowadays for building an audio amplifier from discrete components, unless it is for hi-fi applications.

An example of just what can be done in integrated circuit audio amplifiers is the LM380, a 2-watt audio amplifier. Including the volume control, the complete amplifier uses *six* components. The full circuit is shown in Figure 13.13.

fig **13.13** *a modern integrated circuit audio amplifier; there are only six components and about twenty soldered joints*

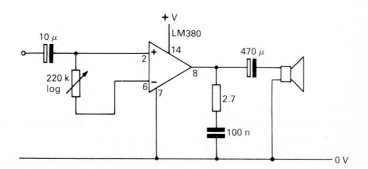

The LM380 has obvious affinities with the op-amp; there are two inputs, inverting and non-inverting, but the output stage is class B. It will deliver 2 W maximum into an 8 Ω speaker when connected to a 20 V supply, but will work with any supply voltage down to a few volts, giving proportionally less output. It also incorporates various safety circuits—it will shut down if it overheats, and the output can be shorted to either supply rail without damage. There is an outwardly similar high-power version, the LM384, that delivers 5 W peak from a 22 V supply.

Figure 13.14 shows the pin connections—the circuit is in the familiar 14-pin DIL pack. The middle six pins are connected to the 0 V supply line, but also provide a heat-sink path for the output transistors. If the printed circuit board is designed with a large area of copper, say 20 cm², connected to these pins, then heat is conducted away from the output transistors and is dissipated by the copper of the printed board itself—a neat design idea.

fig 13.14   *pin connections of the LM380 audio amplifier circuit; the centre three pins on each side are connected to the output transistors and provide a heat-sink path*

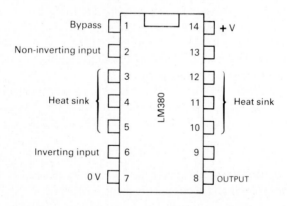

There is a companion IC for the LM380 and LM384, the LM381, which contains most of the components for a stereo preamplifier system.

There are, of course, many different integrated circuit audio amplifiers from many different manufacturers. Power outputs vary from 500 mW to several tens of watts. Heat-dissipation arrangements range from fins built into the device to tags such as are used on power transistors. There is little point in cataloguing them here, except to indicate that the trend is away from complex amplifiers and towards ever simpler systems.

A general exception to this rule is hi-fi systems!

13.5 **HIGH-FIDELITY AUDIO**

There is not even a generally accepted definition of 'hi-fi'. All the standards so far produced have rapidly been thought of as far too low, while in fact the very best equipment now available is probably better than it needs to be. That is to say, the tiny amounts of distortion and 'unfaithful' reproduction are almost certainly far too small for even the most trained human ear to detect. This doesn't stop reviewers of hi-fi equipment thinking they can perceive such differences!

Because integrated circuit audio systems are made for a specific price range, hi-fi amplifiers generally use discrete components in their construction, or at least use op-amps as the 'gain' element in rather complicated designs. The design of hi-fi equipment is very specialised, and well beyond the range of this book, but it is instructive to look at the specification of a good hi-fi amplifier.

Power
The output power should probably be 20 W or more, not because this is a sensible level to use all the time, but because the amplifier is required to reproduce high-energy transient sounds without distortion. Also, modern speakers tend to be inefficient, trading small size and sound quality for electrical efficiency; they require lots of power to drive them to high volumes. Since hi-fi systems are invariably *stereo* (see below), this would be 20 W per channel, a total output power of 40 W. Manufacturers sometimes quote 'peak power', which is the amount of power an amplifier will deliver for a short period. Peak power is about twice the normal (r.m.s.) power.

Distortion
There are various measures of distortion. The most commonly used is *total harmonic distortion* (t.h.d.). A top-quality amplifier would achieve better than 0.01 per cent t.h.d. across the entire audio range.

Frequency response
In the days of valve amplifiers and in the early days of transistor amplifiers it was quite difficult to arrive at a design which successfully amplified signals across the entire audio spectrum. It is not a problem today, and any good amplifier will handle signals from 20 Hz to 20 kHz: 'handle' needs definition in this context—the amplifier has to respond to, say, a 20 Hz signal in the same way as to a 2 kHz signal. Frequency limits are generally quoted to '3 dB down', meaning that the signal inside the quoted limits is no more than 3 dB smaller at the limits than in the middle. This is shown graphically in Figure 13.15.

fig **13.15** *an amplifier frequency-response curve (frequency response is normally quoted at −3 dB limits)*

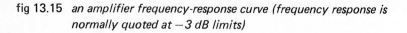
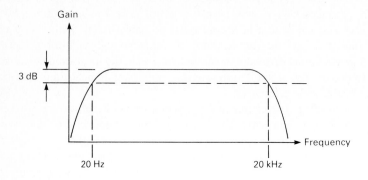

It is not a good idea to have an audio amplifier that amplifies signals to too high a frequency, for ultrasonic frequencies can use up power and heat the output stage to no audible result. Most audio amplifiers are designed so that the frequency response dips dramatically above 20 kHz. In the same way, very low frequencies are not wanted—playing the warps in a record will exercise the speaker cones, but will not contribute to the sound, and may actually detract from it.

Noise

Every amplifier produces some background noise, in the form of a steady 'hiss'. The noise is inherent in the way transistors work and can never be entirely eliminated. However, it is possible to reduce the noise level to a practically inaudible −70 dB, i.e. 70 dB (more than 3000 times) lower than the signal level. Hum, generally at mains frequency, and if not due to problems in interconnections between units, is always due to poor/cheap design. A hi-fi amplifier should have no discernible hum, even with the volume turned fully up.

13.6 STEREO SYSTEMS

The human hearing system is capable of locating objects in space by the relative volume of sound received by each ear. Thus a speaker placed on each side of the listener will be able to duplicate an entire 'sound stage' between the speakers, by presenting the left and right ears with different sounds. This is the principle of stereo sound, and is universal for hi-fi. It does require two of everything—two speakers driven by two amplifiers, with sound recorded in two separate channels. A system diagram of a stereo record-player is shown in Figure 13.16. The relative loudness of the

fig 13.16 *system diagram for a stereo amplifier*

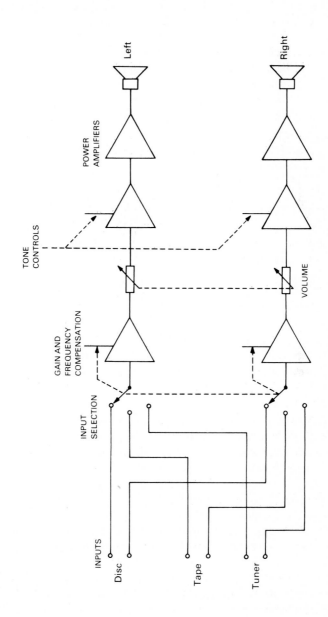

two channels (balance) is controlled by a 'two-gang' (i.e. two on one shaft) potentiometer, and tone controls are also ganged.

Although there is only one pick-up tracking the groove in the record, there are two completely separate channels of information recorded and replayed. Figure 13.17 shows a stylus resting in a groove of a stereo record. The groove is V-shaped, and the two side walls are impressed with separate vibrations for the left and right channels. The stylus is moved as shown in *A* and *B* by the two walls.

fig 13.17   *separate information modulated into the groove of a stereo record; the pick-up cartridge is sensitive to movement in both directions, and movement in plane A or plane B will give separate outputs*

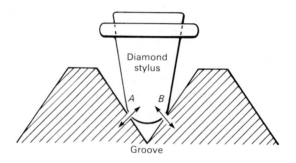

Since the directions of movement are at right angles, the cartridge can mechanically separate the signals and generate separate outputs for the left and right channels of the system. Most modern cartridges are of the *magnetic* (moving-magnet) type, in which the movements of the stylus caused by the record groove are made to move a tiny permanent magnet mounted inside a fine coil of wire. The voltage induced in the coil constitutes the output, at most a few millivolts. There are, of course, two sets of coils, responding to the two channels—although cartridge designs differ from make to make, the principle is the same.

Crystal cartridges and *ceramic* cartridges are used on cheaper record-players, and rely on the *piezoelectric effect.* The motion of the stylus bends a crystal of Rochelle salt, which has the property of generating an electrical voltage across its width when flexed. The voltage is quite high, up to 1 V, but the non-linearity of the output leads to a 'low-fi' output. Ceramic materials can be made that give better linearity but lower output, about 100 mV being typical.

Figure 13.18 illustrates the general layout of a magnetic pick-up cartridge. If the stylus is moved in the direction of the arrows, corresponding

fig **13.18** *internal construction of a moving magnet pick-up cartridge; the moving parts—stylus, lever and magnet—have been shaded for clarity; a pivot, usually of compliant plastic material, holds the moving parts in place while allowing the stylus to track the complex shape of the record groove*

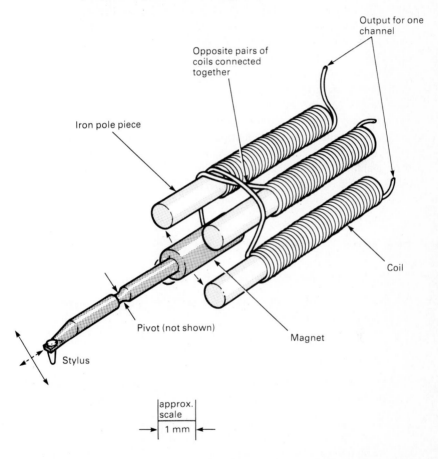

to one channel of the stereo recording, the magnet moves between the upper-left and lower-right pole pieces, inducing a small current in the coils. This is the signal current that is amplified by the record-player. Since the magnet remains centred between the other two pole pieces, no signal is induced in these coils, or at least what small signal there is will be cancelled out.

The situation is reversed when the stylus moves in the other direction; in practice the stylus moves simultaneously in both directions to reproduce the two stereo channels.

13.7 COMPACT DISC DIGITAL AUDIO

The basic principle of sound recording on discs has been unchanged since Thomas Alva Edison invented it towards the end of the nineteenth century. Sound is converted to mechanical vibrations that are used to cut an undulating groove in a suitable material when recording the sound. When replaying, the groove moves a stylus, which converts the mechanical movement back into sound. Huge improvements have been made since Edison's day, but overall, the principle is the same.

1984 (not as bad a year as it might have been, George!) brought the first major departure from Edison's system, if you exclude tape recording. This was the 'compact disc' digital audio system, CD for short. This combines the advantages that the records have over tapes – random access of tracks, speed of mass reproduction, ease of storage – with superlative sound quality. The compact disc system uses digital recording techniques that guarantee near-perfect audio quality. Because of this, the compact disc system is described in the third part of Mastering Electronics, Chapter 18.1, page

QUESTIONS

1. What is the main disadvantage of class A amplifiers?

2. Class B amplifiers often feature internal temperature-compensating circuits. What are these for?

3. Why are integrated-circuit audio amplifiers preferred in most audio equipment, televisions, etc.?

4. What is the maximum power that could be delivered into an $8\,\Omega$ speaker by an amplifier operating from a $20\,V$ power supply?

5. Hi-fi amplifiers are designed so that they will not amplify frequencies substantially higher or lower than the range of frequencies to which the human ear is sensitive. Why?

TAPE-RECORDERS

The principle of tape-recording is straightforward enough, and work on simple magnetic recorders was going on as early as 1900. In the first systems a steel wire was used as the recording medium, but today magnetic tape is universal.

The tape is pulled past a *head* at a constant speed by the drive mechanism. The head consists of a core, made from a magnetic but non-conducting material such as ferrite, or sometimes laminated iron. The core is made as a closed circle, with an extremely narrow gap where the core touches the tape. A coil of insulated wire is wound on the core, shown diagrammatically in Figure 14.1.

fig **14.1** *a tape record/playback head*

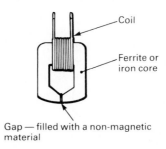

Coil

Ferrite or iron core

Gap — filled with a non-magnetic material

The tape itself consists of a flexible (but non-stretch) plastic base, coated on one side with a magnetic powder. Ferrite (an oxide of iron) is commonly used, but other materials such as chromium dioxide are now used as well. The particle size is extremely small.

If an alternating current is passed through the coil in the head, a magnetic field will be induced in the gap in the core, and the field will vary in proportion to the current flowing. If the tape is moving past the head at the time, the varying magnetic field will cause a pattern corresponding to the alternating current to be recorded on the tape in the form of changes in patterns of permanent magnetism. Figure 14.2 illustrates this. The alternating current applied to the head can be obtained from the output of an amplifier, and the input to the amplifier can be via a microphone. This will permit recording speech on the tape, in the form of magnetic patterns.

fig 14.2 *during recording, the head produces patterns of magnetisation on the magnetic tape*

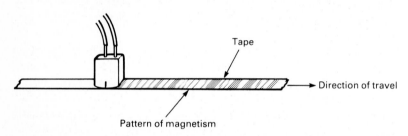

To replay the tape, the head is connected to the amplifier *input*. As the tape is pulled over the head, the permanent magnetic patterns previously recorded on it will induce a voltage in the coil that is closely similar to the speech pattern. The amplifier's output is connected to a speaker, and the system will replay the original speech—more or less. A simple system diagram for a tape-recorder is given in Figure 14.3. The switch is shown in the 'playback' position (marked *P*) and the head feeds the amplifier, which drives the speaker. In the 'record' position (marked *R*) a socket—for a microphone—is connected to the amplifier input, and the output is fed to the head. Although a moderately large signal, typically a few volts, is needed to record a signal on the tape, the output from the head during playback is small. One channel of a stereo cassette recorder head will give at most a few hundred microvolts of signal.

fig 14.3 *system diagram for a simple tape-recorder*

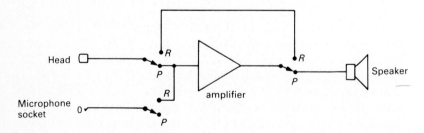

The tape can be erased by applying a powerful magnetic field to saturate it and destroy the previous recording. In low-cost portable recorders a small permanent magnet is used for this purpose, swung into contact with the tape when the 'record' button is pressed. Erasing a tape in this way leaves a residual noise (hiss) on the tape, so in higher-quality recorders the tape is erased with a powerful high frequency alternating magnetic field, applied with a special *erase head*. The erase head is similar to a record/ playback head, but has a wider gap and, usually, a larger coil—the object is to deliver a high power field to the tape, not to record a signal. Tapes erased by demagnetising them are quieter, with less background noise, than tapes erased with a magnet.

Unfortunately a tape-recorder based on such a simple circuit as the one shown in Figure 14.3 sounds dreadful. Replay is satisfactory, but substantial distortion is introduced during recording. The problem is caused by a basic physical fact, that induced magnetism is not linear. If there is a very small signal to be recorded, a small degree of permanent magnetism is left in the tape—this is called the *remanent magnetism*. Doubling the field will not double the remanent magnetism; at low levels and at high levels the remanent magnetism changes more slowly than the inducing field. Figure 14.4 shows a graph comparing the shape of the 'curve' for a steadily increasing field strength with the curve for the remanent magnetism which would result. The middle of the right-hand curve is flat, but distortion will occur if the signal is recorded near the top or bottom of the characteristic.

fig 14.4 *the remanent magnetism in the tape does not bear a linear relationship to the magnetising force*

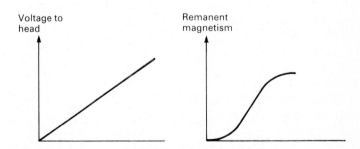

When an audio signal, which is alternating, is fed to the record head, the direction of magnetism reverses every time the waveform crosses the zero line. Figure 14.5 shows what happens when a simple sinewave is used.

This means that the sinewave will be distorted, as shown in the diagram, every time it approaches the zero line. Avoiding the distortion at the 'high' end of the characteristic is easy—it is merely necessary to ensure that the

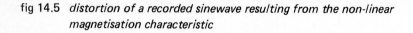

fig 14.5 *distortion of a recorded sinewave resulting from the non-linear magnetisation characteristic*

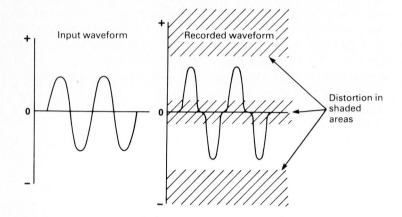

signal is not too loud. But the problem near zero is harder to cure. The method is in fact to mix the audio signal with a constant high frequency note, a sinewave at 40 kHz or more, well beyond the audible range. This technique, known as *ultrasonic bias* (or sometimes, less correctly, 'supersonic bias'), ensures that the audio waveform is undistorted, as it is all recorded on the linear part of the magnetic curve. Figure 14.6 illustrates how it works. In Figure 14.6*a* the original waveform produces a distorted recording on the tape. In Figure 14.6*b* the same waveform, when mixed with an ultrasonic bias frequency, produces a mixed waveform on the tape—but notice that the top and bottom limits of the mixed signal constitute the audio signal, and they are well away from the zero point. The distortion is still there, of course, but it only distorts the bias signal, which isn't wanted anyway.

When the tape is replayed, the bias frequency is removed with a simple filter (often a single capacitor across the signal path) and the original waveform is recovered, undistorted, for amplification.

Because different types of tape (ferric, chromium, metal) have different magnetic characteristics, different levels of bias are required for each. Hi-fi tape-recorders are therefore equipped with bias-select controls which enable the user to produce undistorted recordings on all three main types of tape.

Volume, tone and sometimes automatic level control for recording all adds to the complexity of even inexpensive recorders. In many, the component count is reduced by the user of purpose-built integrated circuits which incorporate several different functions on one chip. The best-quality recorders, as with hi-fi amplifiers, are made using discrete components.

fig 14.6   *comparing recording systems with and without bias; the addition*
of ultrasonic bias provides a means of recording an undistorted
audio waveform on magnetic tape

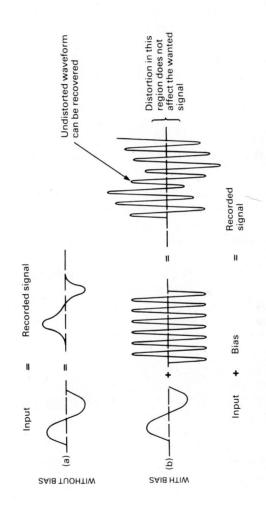

Some recorders have completely separate record and playback systems, with separate record and playback heads. This permits listening to the recorded signal a fraction of a second after it has been recorded on the tape; at the flick of a switch it is possible to compare the incoming signal with a recorded signal.

Two main types of tape-recorder, *cassette* and *reel-to-reel*, are in use. Recording studios use a whole range of formats, all reel-to-reel, but the standard 'domestic' reel-to-reel recorder uses $\frac{1}{4}$-inch wide tape. The tape carries four tracks, and the machine records two in each direction (stereo), and may have the facility to record all four tracks separately. Running speeds of $3\frac{3}{4}$ or $7\frac{1}{2}$ inches per second are usual. Reel-to-reel recorders have now been superseded in most domestic applications by the cassette recorder, which uses a tape $\frac{1}{8}$ inch wide, running at $1\frac{7}{8}$ inches per second. Like the reel-to-reel format, the tape has four tracks recorded on it, though they are arranged differently. Figure 14.7 shows a comparison between the reel-to-reel tape and the cassette tape.

fig **14.7** *track positions on reel-to-reel and cassette tapes; note that the reel-to-reel tape records both pairs of tracks with a gap between them, which simplifies head design to reduce 'crosstalk'; cassette tape-recorders record tracks adjacent to one another, so that they can be replayed together by a mono machine, easing the compatibility problem at the expense of worsening crosstalk*

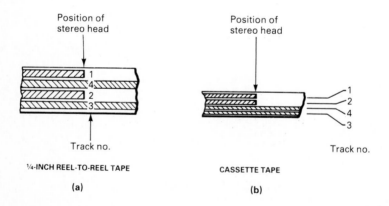

Technical advances have made it possible to obtain true hi-fi results from cassette tapes, which are both much cheaper and much more convenient than reel-to-reel tapes. Improvements in the tapes themselves have been a major factor, as has the improvement in electronics. And the introduction of Dolby® noise-reduction systems have lowered background noise to better than −70 dB, inaudible in practice. Dolby® noise-reduction techniques are very complicated, but in principle work by boosting the high frequencies when recording low-level sounds. At higher levels the high frequency boost is reduced, to avoid overloading the tape. During playback, a decoder system reverses the process, so high frequencies are attenuated to restore the correct balance to the recording—and along with them the background noise is attenuated as well. At higher recording levels, there is less improvement in background noise, but this is unimportant as the high-level sound masks the noise anyway. There are various implementations of the Dolby® system, the most common being *Dolby B* and *Dolby C*. Extraordinary improvements in background noise result, the two systems providing a noise reduction (in hiss, particularly) of 10 dB and 20 dB respectively.

14.1 TAPE DRIVE SYSTEMS

Along with an improvement in tapes and electronics has gone an improvement in the drive system used to transport the tape past the head. It should be evident that the tape must travel very smoothly past the head; any speed variations will alter the pitch of the playback. Slow changes in drive speed are known as 'wow', and rapid changes (which sound like someone gargling) are known as 'flutter'.

Figure 14.8 illustrates a basic cassette recorder drive, omitting the extra pulleys and belts to provide fast forward and rewind.

fig **14.8** *a schematic diagram of a cassette tape drive mechanism (for clarity, this illustration omits the extra drive necessary to rotate the take-up reel of the cassette)*

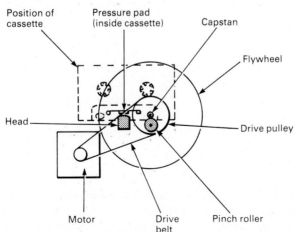

The motor, in battery-powered recorders, will often have an electronic speed control. The basic system for this is to have a generator fixed to the motor shaft, the generator producing a voltage that is more or less proportional to the speed of rotation of the shaft. An electronic circuit (op-amp?) compares the generator voltage with a fixed reference voltage derived from a Zener diode. If the motor is going too fast, the generator voltage will be high, and the circuits will reduce the voltage to the motor. If, on the other hand, the generator voltage is low, the circuits will increase the motor voltage. Very accurate speed control can be obtained in this way, and the motor's running speed will be substantially independent of the battery voltage. This form of speed control is referred to as *servo control*. A system diagram is given in Figure 14.9.

fig **14.9** *electronic speed control of a motor using a feedback servo system*

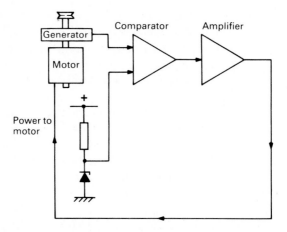

A heavy flywheel smooths out vibration from the motor, and provides momentum to ensure that minor 'stickiness' in the cassette will not affect the running of the tape. The tape is actually transported by a *capstan*, against which the tape is held by a rubber *pinch roller*. The capstan is always of small diameter, so that the flywheel can rotate fairly quickly, and will always have a big leverage advantage against the tape.

14.2 VIDEO TAPE-RECORDING

The signal that provides the picture in a television can be recorded on tape in much the same way as an audio signal. It is, however, technically much more difficult to record a video signal because of the very large *bandwidth*:

the recorder must reproduce very high and very low frequencies. This rules out the recording system used for audio, for a record head can only record a range of frequencies covering about 10 octaves (each octave doubles the frequency, so 10 octaves higher than 20 Hz is 10.24 kHz). The system used is *frequency modulation* (see Chapter 15), which allows the required bandwidth of 18 octaves or more.

The second problem is reproducing the highest frequencies, around 5 MHz, at all. To record a single cycle of a waveform, a length of tape at least three times as long as the width of the record head gap must pass the head. Even with the finest heads, the required tape speed is very high indeed. Early video recorders, which operated like very large audio tape-recorders, used tape speeds of up to 9 metres a second!

The solution to this problem is the use of rotating heads. Two heads are mounted in a cylindrical drum, and the tape is wound round the drum at an angle. While the tape is pulled over the drum by the tape drive capstan, the drum spins at high speed in the opposite direction, the heads whizzing diagonally across about a foot of tape. This design is known as *helical scanning*.

Each complete scan across the tape is equal in time to one television field, which has the advantage that if the tape drive is stopped, but not the rotating drum, a still picture will be left on the TV screen.

An idea of the way helical scanning works can be obtained from Figure 14.10. The tracks are very narrow indeed, for the more that can be crammed in, the longer the tape will play. On a domestic machine, the width of the tracks may be only about forty *thousandths* of a millimetre! The actual speed of tape through the machine is between 15 and 30 mm per second, according to the type of machine.

QUESTIONS

1. What is meant by 'bias' in relation to tape-recording?

2. Describe the function in a tape-recorder of (i) the erase head, (ii) the record/playback head, (iii) the flywheel.

3. Describe the symptoms you would expect in the case of failures in the following parts of a tape-recorder: (i) electronic speed control, (ii) ultrasonic bias, (iii) audio amplifier.

4. How can a video tape-recorder produce a still picture when the tape drive is stopped?

5. Why are noise-reduction systems more important in cassette tape-recorders than in reel-to-reel recorders?

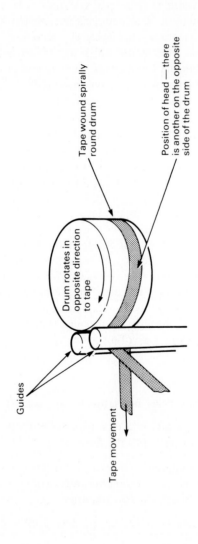

fig 14.10 *helical scanning of a video tape provides a very high writing speed, with a relatively slow tape transport speed*

RADIO AND TELEVISION

Radio transmission and reception was perhaps one of the earliest applications of electronics, and is—so far—the application that has made the greatest impact on society. Oddly, we can use radio, predict its properties and design circuits that work very efficiently, but we know little about the real nature of radio. Ask an electronics engineer what radio is, and the answer will be a confident, 'Electromagnetic waves.' Ask a physicist what electromagnetic waves are, and he will begin to hedge, or he will tell you that really we don't know. We do know that electromagnetic radiation is a form of energy, and that it behaves as if it is propagated as waves. The model becomes more of a model and less like reality when we discover that radio travels through a vacuum. How can there be waves in a vacuum? Perhaps in the future, theoretical physics will give us an answer. In the meantime, we use radio, describe it mathematically, and design and use electronic circuits that function happily despite our underlying ignorance.

Possibly the hardest concept to grasp in radio is the way in which circuits can be made that broadcast radio waves into their surroundings.

We begin with the *LC* tuned circuit, given in Figure 10.20 (p. 138). You will remember how the circuit, when oscillating, stored all the available energy, first in the capacitor and then in the magnetic field associated with the inductor, reversing the situation with each swing of oscillation. Consider the way in which energy is stored in the capacitor (Chapter 3)— it is in the form of an electric field, whatever that may be! All capacitors that we use in electronic circuits develop an electric field between two parallel and very closely spaced plates. Moving the plates apart has two effects. First, it decreases the capacitance (the principle of variable capacitors); and second it makes the electric field occupy a larger volume of space. We could make a capacitor with two plates a metre apart, which would give a tiny value of capacitance unless the plates were *very* large indeed.

Operating this cumbersome arrangement in an *LC* oscillator circuit would yield an unexpected result, for not all the energy put in to the circuit could be accounted for by waste heat emitted by the coil windings, etc. Some of the energy has escaped—leaked away, if you like—from the area between the plates (which are rather a long way apart). This escaped energy is, of course, electromagnetic or 'radio' energy, and has been broadcast away from the circuit. In order to make a radio transmitter, we need a circuit that will give the maximum possible 'leakage' of energy from the *LC* oscillator.

As you might expect from the hypothetical experiment outlined above, what is needed is a capacitor with an enormous gap between the plates, and designed in such a way that the largest possible amount of radio is 'leaked'.

We can, in fact, use the largest possible plate for one of the plates of the capacitor—we can use the earth! The other 'plate' has to be rather smaller, and it is convenient to use a single, long wire; this has the advantage that it can be made an excellent 'leaker' of radio energy. The earth and wire as a capacitor is shown in Figure 15.1. The picture is a familiar one, that of a radio aerial.

fig 15.1 *an aerial*

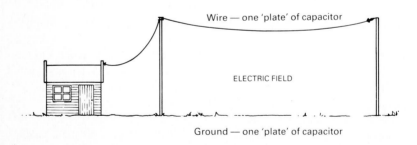

Wire — one 'plate' of capacitor

ELECTRIC FIELD

Ground — one 'plate' of capacitor

It also turns out that there is an optimum length for the aerial wire, if the 'leakage', which I shall now call 'broadcast energy', is to be maximised. Conveniently, it is when the aerial is half, or one-quarter, of the *wavelength*. The term 'wavelength' refers to the physical distance between complete cycles of the broadcast wave. Radio waves travel at the speed of light, which is about 300 000 km per second. If the radio waves were produced at the highly unlikely rate of one a second, the wavelength would be 300 000 km. Two a second, and the wavelength would be 150 000 km, still a little on the long side for a convenient aerial to be one quarter of the wavelength!

Radio is broadcast on *much* higher frequencies than this, but the formula for working out the wavelength is the same:

$$\lambda = \frac{v}{f}$$

(where λ is the Greek letter 'lambda', which is conventionally used to represent wavelength).

At a radio frequency of 10 MHz, the wavelength would be

$$\lambda = \frac{300\,000\,000}{10\,000\,000}\ \text{metres}$$

$$\lambda = 30\ \text{m}$$

A convenient quarter-wavelength aerial would be 7.5 m, not too bad, but unsuitable for portable equipment. For portable radio transmitters, we can exchange convenience for efficiency, and use a 1/8 wavelength or a 1/16 wavelength aerial.

The length of the aerial will affect the capacitance (refer to Figure 15.1 and you will see why this happens) and this will affect the frequency of oscillation. For a transmitter circuit to work properly, the *LC* circuit must be 'tuned' to the transmission frequency, and the aerial must be the correct length. Radio frequencies between 30 kHz and 20 GHz or so are used. Figure 15.2 overleaf illustrates the different frequency 'bands' and the names given them.

15.1 RADIO TRANSMITTERS

We can now look at a practical transmitter circuit, operating in the high frequency band at 27 MHz. The choice of frequency is not arbitrary, for this is the model radio control frequency, and in the UK and many other countries it is legal to operate a low power transmitter at this frequency without any form of license. Many readers can therefore build and operate this circuit—provided it is properly tuned—without breaking the law.

In order to ensure that they are operating on a legal frequency, model control transmitters have to be *crystal controlled*, i.e. the frequency of oscillation must be controlled by a quartz crystal oscillator, to ensure accuracy and stability. A frequency close to the middle of the model control band is 27.095 MHz, and crystals at this frequency are readily available from good electronic component or model shops.

The crystal-controlled oscillator circuit in Figure 10.24 (p. 140) is a good starting-point. This oscillator provides the necessary radio frequency sinewave output, which is then amplified by a simple transistor amplifier (see Figure 15.3 on page 199).

fig 15.2 *the radio spectrum*

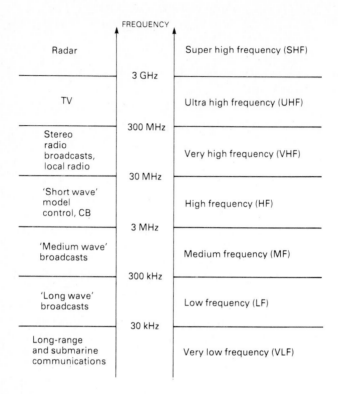

Note the use of a tuned circuit for the transistor amplifier, something which is fairly common in radio circuits. The use of a tuned collector load improves the efficiency, for the impedance of the load is highest at the 'wanted' frequency, giving the best output and optimum bias. In this case the frequency of the resonant circuit is about 27.1 MHz.

The specification of inductors is always problematic, because there are many different ways of making, say, a 10 mH inductance, and not all are suitable for all applications. In practical circuits it is usual to give constructional details of any 'purpose-made' inductors, to ensure that the right design is used for the circuit. Fortunately for us, the amounts of inductance required in high frequency radio circuits are generally quite small, and suitable inductors are easily made with a few turns of enamel-insulated wire over a small ferrite core (or no core, in which case the inductor is called 'air-cored').

To give a reasonable range with a typical model control receiver, our transmitter needs an output of 200–500 mW. A third, *power stage*, is used to give the high output current needed.

fig **15.3** *a crystal-controlled oscillator circuit, followed by a tuned ampli-*
fier stage; this practical circuit can form the basis of a transmitter

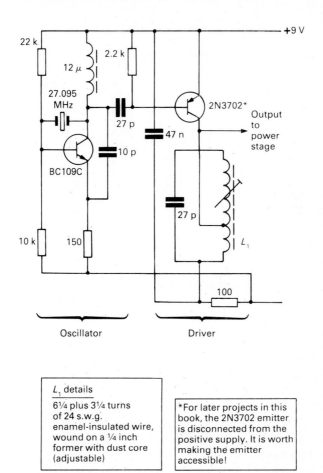

L₁ details

6¼ plus 3¼ turns
of 24 s.w.g.
enamel-insulated wire,
wound on a ¼ inch
former with dust core
(adjustable)

*For later projects in this
book, the 2N3702 emitter
is disconnected from the
positive supply. It is worth
making the emitter
accessible!

The power stage is very similar to the preceding stage, using a very
simple transistor amplifier. However, a larger, high-power transistor is
needed, and the collector load (and thus the bias resistor) are much smaller
in value to provide the required current. The circuit is given in Figure 15.4.

Notice the resistor values associated with the power stage, and notice
also the small capacitor (100 pF) used for coupling the stage to the one
preceding it: the high frequency means that a small capacitor is able to
pass sufficient current.

The last part of the circuit, following the power stage, is the output
tuning circuit. There are various forms, but it is usual to have two tuned

fig **15.4** *the output stage of a small transmitter; this can be driven directly by the circuit in Figure 15.3*

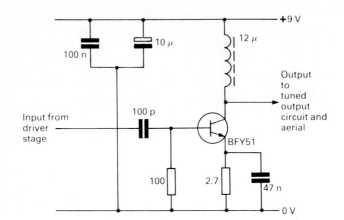

Power amplifier

Power transistor should be fitted with clip-on heat radiator

circuits—both *LC* circuits—to improve efficiency. The circuit shown in Figure 15.5 is typical.

There are a number of noteworthy points about this circuit. First, the inductors are generally air-cored (again, for efficiency) but may be wound on ferrite cores to increase the inductance if the transmitter has to be small. Second, the capacitors are variable, so that the circuit can be tuned accurately—the *LC* circuits have to be accurately 'peaked' (i.e. tuned to the exact frequency) for the transmitter to work properly. The same is true of any other tuned circuits, such as the collector load of the stage driving the power stage.

The capacitor that couples the aerial itself to the circuit is there for safety reasons, and does not play any part in the circuit's functioning. If the aerial is accidentally shorted to the transmitter casing (which is earthed to the 0 V power line) this would short-circuit the output transistor and could damage the driver transistor; but the 10 nF capacitor prevents this.

In practice, the circuit can be built on a printed board, and layout is uncritical. Matrix board is not suitable, since the strips of copper act as unwanted capacitors—at high frequencies a few picofarads can be important. The coils (in this design) are intended to be mounted facing the same way (i.e. with the same axis).

fig **15.5** *output tuned circuits for a 27 MHz transmitter; this circuit can be driven directly by the output amplifier in Figure 15.4*

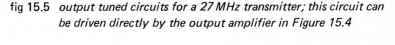

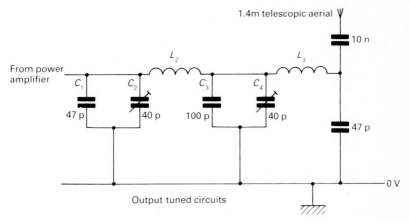

The circuit shown in Figures 15.4 and 15.5 is entirely practical, and can be made as a construction project. The crystal *must* be suitable for operating on the model control band, between 26.995 and 27.245 MHz in the UK.

The transmitter can be built, but we need some kind of detector to check that it is working, and to allow us to tune the *LC* circuits. A suitable detector is shown in Figure 15.6. This is very simple, and will react to the 27 MHz radio frequency when held a few centimetres from the transmitter aerial.

The radio signal induces a small current in the wire, and this flows through the meter. Without the diode, an alternating current would flow, which would not perceptibly affect the meter—the sluggish needle is unable to swing to and fro 27 million times a second! The diode *rectifies* the signal, and makes the current flow through the meter unidirectional.

fig 15.6 *an RF detector suitable for use with a low-power 27 MHz transmitter*

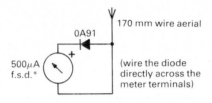

*Full-scale deflection.

The meter can measure the current, and responds to the radio signal. Such a device is referred to as a *detector*, or *RF meter* (radio frequency meter). The diode specified is germanium, not silicon. Thermal effects and leakage are unimportant in this application, but forward voltage drop *is*—and the germanium diode is better than silicon in this respect (germanium semiconductors still have their place).

The RF meter can be used to tune the transmitter. Connect the power supply to the transmitter, and bring the RF meter close up to the fully extended aerial. Adjust the core of L_1 with a plastic screwdriver (not a metal one, which would affect the inductance). At one point, the meter will show a small deflection. Adjust the core for maximum deflection of the meter. Now adjust the output capacitors to improve the deflection. You may need to move the RF meter further away from the aerial, but keep on adjusting L_1, C_2 and C_4 until no further improvement can be obtained. This is the way to ensure that all the tuned circuits are set as accurately as possible to 27.095 MHz.

The final adjustment should be made with the transmitter in its metal case, held in the hand. The reason is that the connection to earth is made through the operator, who will be holding the transmitter and will have his feet touching the ground.

15.2 MODULATION AND DEMODULATION

Although this transmitter can send out quite a powerful signal, the signal itself carries no information. The RF meter can detect the presence or absence of the signal, but if the radio waves are to carry useful information, such as speech or control signals, the system needs additions.

There are two common methods of adding information to the radio signal, which is called the *carrier*. Each involves changing the carrier slightly, and both systems are in common use. The first, and most obvious, is *amplitude modulation* (AM). Amplitude modulation involves nothing more complicated than changing the power, or amplitude, of the carrier in sympathy with the modulating signal. This is clearly illustrated in Figure 15.7, which shows the carrier being modulated with an audio-frequency sinewave.

fig **15.7** *amplitude modulation*

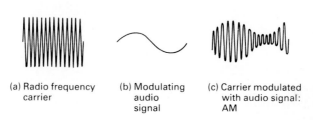

(a) Radio frequency carrier   (b) Modulating audio signal   (c) Carrier modulated with audio signal: AM

AM has the additional advantage that it is easy to recover, or *detect* at the receiver. Assuming that the signal received by the receiver is roughly the same as that shown in Figure 15.7c, we cannot simply feed the output into a speaker. The output is, at audio frequencies, symmetrical so that any increase in positive signal is exactly balanced by a similar increase in negative signal and the result is zero.

Fortunately, the audio signal can be extracted simply by rectifying the receiver's output, to make it asymmetrical. The carrier is then removed with a small capacitor. The circuit is shown in Figure 15.8, along with the waveforms associated with it. Comparison with Figure 15.7 shows how the modulating waveform is recovered more or less unchanged.

The second method is known as *frequency modulation* (FM). Instead of changing the amplitude of the carrier in sympathy with the modulating waveform, the frequency is shifted a little higher or a little lower. The amount of frequency shift is very small. Frequency modulation of a carrier is illustrated in Figure 15.9; compare it with the same diagram for amplitude modulation, Figure 15.7.

Detection of the FM signal is much more complicated than detection of AM signals. Various circuits have been developed, and up to the beginning of the 'integrated-circuit era' a circuit known as a *ratio detector* was most commonly used to recover the modulating signal. The circuit is given in Figure 15.10. The functioning of this circuit is quite complicated, and it is now little used.

Nowadays a circuit called *phase-locked loop* is often used. A system diagram is shown in Figure 15.11 (p. 206). The circuit operates as follows. The

fig **15.8** *detector circuit designed to recover an audio waveform from an AM radio signal*

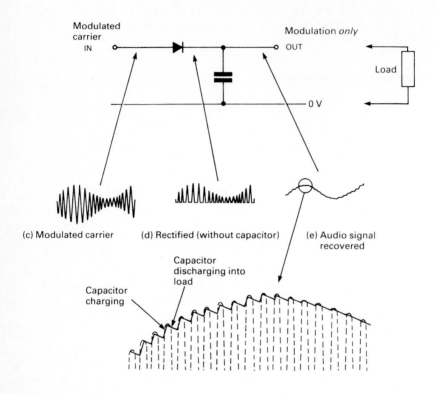

(c) Modulated carrier   (d) Rectified (without capacitor)   (e) Audio signal recovered

fig **15.9** *frequency modulation*

(a) Radio frequency

(b) Modulating audio signal

(c) Carrier modulated with audio signal: FM

fig **15.10** *a ratio detector circuit used for recovering the audio signal from an FM transmission*

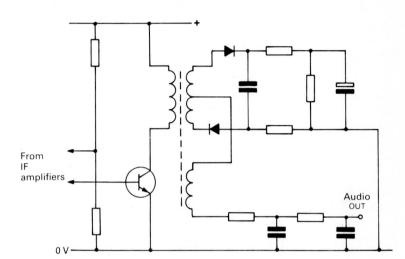

frequency-modulated signal from the receiver is fed into one input of the *phase detector*. The other input of the phase detector is connected to the output of a *voltage-controlled oscillator* (VCO), working at the same frequency as the carrier. The phase detector compares the phases of the two signals—see Figure 15.12—and, if the two signals are out of phase, produces a positive or negative output according to the direction of phase error.

The output is fed to the VCO, which changes frequency in such a way as to move its output signal back into phase with the incoming frequency-modulated carrier. The VCO therefore 'tracks' the frequency changes of the carrier, continuously altering its own frequency to correspond with the incoming frequency modulation. The VCO 'locks' on to the carrier, through a feedback loop (hence 'phase-locked loop'). Now look at the control voltage applied to the VCO by the phase detector. A few moments' thought will show that this control voltage accurately reflects the changes in frequency of the incoming signal—it has to, in order to make the VCO change frequency to track the modulation. The control voltage therefore reproduces very accurately the modulating signal, and is simply amplified to provide an audio output.

Another, very important, advantage of the phase-locked loop is that it not only tracks the modulation but will, within a certain range, track overall changes in frequency of the carrier, or will compensate automatically for any 'drift' in its own operating parameters. This feature enables

fig 15.11 *system diagram for the phase-locked loop system of detecting a frequency-modulated signal*

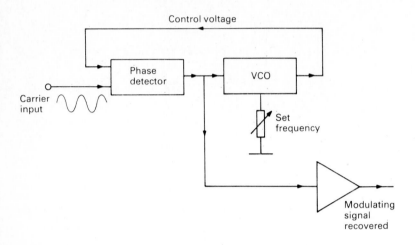

fig 15.12 *signals in and out of phase*

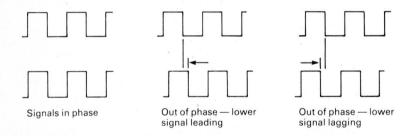

Signals in phase   Out of phase — lower   Out of phase — lower
                            signal leading   signal lagging

the detector to lock on to a signal when the receiver is tuned sufficiently nearly to the transmitter frequency. You will notice this feature working in almost all modern FM radio receivers.

Phase-locked loop integrated circuits specifically designed for radio receivers will usually incorporate other 'convenience' features. These include an output to operate an indicator (usually a LED) when the circuit is locked on to a signal, a means of turning off the feedback to unlock the circuit, and a means of changing the 'capture range' of the circuit (i.e. the amount of deviation from the selected frequency before the circuit unlocks).

It is obvious that the phase-locked loop is actually a rather complicated circuit. Its internal design is certainly complex, but this is unimportant in the context of ICs. The IC is treated as a single component by manufacturers of radios, and as a 14--pin DIL pack (typical packaging for this component) it simplifies and cheapens assembly of an FM radio receiver.

To return to the practical transmitter circuit of Figures 15.3–15.5, the output can be amplitude-modulated by varying the gain of the driver stage with an applied signal. A simple way of doing this is to remove the base connection of the driver transistor from the positive supply rail and connect it to the output of an amplifier or transistor switch that can swing between 0 V and the positive supply voltage. The output of the transmitter will vary in more or less linear fashion with the voltage. There are, of course, many different circuits for modulating a transmitter, all achieving the same result.

If the demonstration transmitter has been constructed, a modulator designed for model control can be added. Details are given in Chapter 22, since the pulse-width modulation system used involves digital systems. In the UK at least, it is illegal to add an audio modulator to this transmitter. Figure 15.13 illustrates the connection for the modulator—refer also to Figure 15.3.

fig **15.13** *modification to the transmitter circuit to enable modulation to be applied*

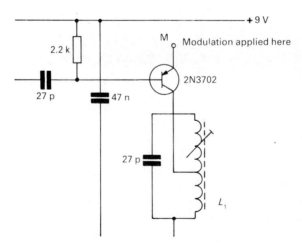

Frequency-modulation techniques are beyond the scope of this book, but a simple FM modulator uses a varicap diode, in parallel with the oscillator crystal, to obtain small changes in transmission frequency according to an applied modulating voltage.

15.3 **RADIO RECEIVERS**

The simplest radio receiver consists of just a tuned circuit, a diode, a capacitor, a pair of headphones, and as much aerial wire as possible! This is the 'crystal set' of the 1920s, and is shown in Figure 15.14.

fig 15.14 *a 'crystal set'*

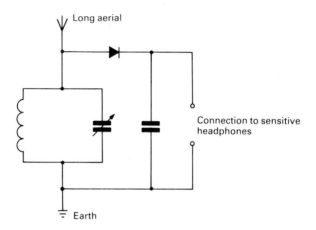

The electric field from the transmitter induces a tiny current in the aerial wire—which needs to be a few tens of metres long—and the *LC* resonant circuit selects the required frequency. It does this because the *LC* circuit has low impedance at frequencies other than the resonant frequency, and this 'shorts out' unwanted radio transmissions. At the resonant frequency the *LC* circuit has a high impedance, so the transmission at the selected frequency appears across the circuit as a small radio frequency alternating voltage.

This is rectified by the diode to recover the audio signal, using a diode with a low forward voltage drop (germanium). The capacitor removes the RF component—Figure 15.8—and a pair of sensitive high-impedance headphones produce a (just) audible output.

The output can, of course, be amplified. However, results from this circuit are still unsatisfactory, chiefly because the tuning cannot separate one radio station from the next. Often, several stations will be received together.

There is a need to make the receiver more *selective*. As we saw earlier, when discussing transmitters, the use of a tuned *LC* circuit as a transistor amplifier's collector load—instead of a resistor—makes the circuit selective, amplifying the wanted frequency much more than other frequencies. If, therefore, a tuned transistor amplifier of this type follows the tuned aerial circuit, selectivity is enormously improved. The *sensitivity* of the receiver is also better, as the radio frequency signal is amplified before it reaches the detector diode. A receiver incorporating a single tuned RF amplifier is shown in Figure 15.15.

fig **15.15** *TRF receiver: only moderately efficient*

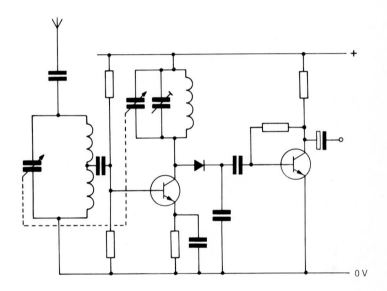

Receivers of this type can be made to work quite efficiently, and are very simple to construct. There is, in fact, an integrated circuit version of this *tuned radio frequency* (TRF) receiver, the ZN414. The ZN414 makes an ideal construction project as it is by far the simplest radio receiver to build. Unfortunately, it will not operate at frequencies higher than 3 MHz, so cannot be used with the model control transmitter. It can, however, be used very effectively with the amplifiers illustrated in Figure 13.9 (p. 175) and 13.3 (p. 177) and will connect directly to them. The circuit for the ZN414 radio is shown in Figure 15.16. The two forward-biased diodes appear to short-circuit the power supply to the ZN414, but this IC requires only a 1.5 V supply, and the combined forward voltage drop of the diodes (0.7 V + 0.7 V = 1.4 V) produces a reference voltage that is near enough right. The circuit takes only 300–500 μA from the supply!

There is no external aerial. With the sensitivity of the receiver vastly increased compared with the simple crystal set, the inductor itself can pick

fig 15.16   *circuit of a modern IC TRF receiver; this provides a low-level
output suitable for feeding to an audio amplifier—this is a
practical circuit and will work well with the amplifier designs
of Figures 13.9 and 13.13*

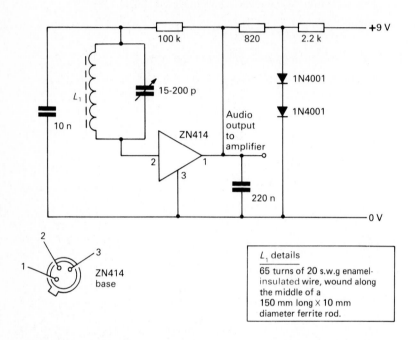

up a sufficiently large signal to work the receiver. A large ferrite core a few centimetres long is used for the aerial, with the coils of the inductor of the tuned aerial circuit wound on it. The whole assembly is called a *ferrite rod aerial.* The aerial is quite directional and can be used to tune out unwanted signals.

Although capable of a respectable performance, the TRF has definite limitations and the ZN414 may represent a development approaching the ultimate for this type of design.

A much better bet from the point of view of selectivity in particular is a receiver consisting of a number of *tuned* amplifier stages. It turns out that this is difficult to build, and even if built gives a poor audio quality. If an audio frequency signal is used to modulate a radio frequency carrier, the result will not be a signal at one frequency alone, but a signal that has components at a whole range of frequencies. The lowest will be the carrier frequency *minus* the highest audio frequency and the highest will be the carrier frequency *plus* the highest audio frequency. A medium wave radio station transmitting at 500 kHz, with an audio signal between 40 and

9000 Hz, would therefore actually be broadcasting on frequencies between 500 − 9 = 491 kHz and 500 + 9 = 509 kHz. And the radio receiver would need to respond to all frequencies between these two figures equally, and ideally not respond at all to frequencies above 509 kHz or below 491 kHz. This is not possible to attain using a series of tuned circuits, for the combined effect is too selective at the chosen frequency.

Much more serious deficiencies come to light when we try to tune the receiver. With several stages, each with its own tuned circuit, we have to contrive a means of tuning *every LC* circuit simultaneously when changing frequency (station). Such a design would be very cumbersome, though a few were made in the 1930s.

There is a solution, one that is employed by all commercially made radio receivers. The design is called the *superheterodyne* (superhet) receiver, and a system diagram is given in Figure 15.17.

fig 15.17 *system diagram for a superhet radio receiver*

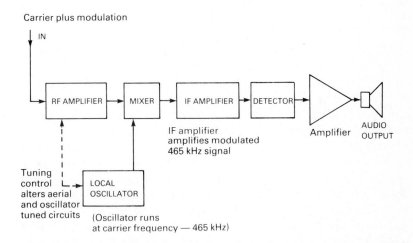

The incoming signal from the radio frequency amplifier (the RF amplifier may be missing in very simple receivers) is *mixed* with a signal from an oscillator, known as the *local oscillator*. The oscillator is tuned to a slightly lower frequency than the selected radio frequency. In just the same way that the transmitter produced sum and difference frequencies above and below the carrier frequency, so the mixer produces an output that is the difference between the frequencies of the received radio signal and the output of the local oscillator. The oscillator is designed, in most AM receivers, to run precisely 465 kHz below that of the received carrier. A ganged (double) variable capacitor is used to alter the tuning of the aerial circuit and the oscillator simultaneously, always keeping them 465 kHz apart. The output of the mixer is therefore always a 'carrier' at 465 kHz, and this carrier is still modulated with the original audio signal.

Now it is possible to use two or three tuned amplifier stages if necessary. The lower frequency makes design much easier, and the tuning of the stages can be arranged to give a selectivity response close to the ideal. Once the signal has been amplified sufficiently, a detector diode (or, for FM, a suitable discriminator circuit) can follow, with audio amplication last in the line.

The frequency resulting from the carrier and local oscillator inputs to the mixer is called the *intermediate frequency* (IF), and the stages that follow are called *intermediate frequency amplifiers*. They invariably have tuned collector loads: tuned, of course, to the intermediate frequency. The intermediate frequency need not be 465 kHz. In model control receivers it is usually 455 kHz, and in VHF radio receivers it is 10.7 MHz. The range of intermediate frequencies used is limited, not for any technical reasons but for the purely practical reason that manufacturers produce IF tuned transformers, known as *intermediate frequency transformers* (IFTs) that are preset to the intermediate frequency, and it is convenient to make just a few types. IF amplifier stages invariably use transformer coupling, for the tuned collector load simply needs a secondary winding to make it into a coupling transformer as well, which saves components and makes for an efficient circuit. A transformer-coupled amplifier stage was shown in Figure 10.11 (p. 128).

Receivers built with discrete components generally combined the local oscillator and mixer into a single transistor stage by means of a circuit like the one illustrated in Figure 15.18. The inductor of the tuned circuit L_1 $C_1 + C_2$ is wound on the ferrite rod aerial, along with a coupling winding L_2 which works as a transformer and provides a signal current for the transistor. The transistor, as well as working as an RF amplifier, acts as an oscillator, L_3 and L_4 providing feedback from the collector to the emitter to sustain oscillation. The frequency of the oscillator is determined by the tuned circuit L_5 $C_3 + C_4 + C_5$ and the tuned IFT selects the intermediate frequency to pass on to the next stage (the IFT has a low impedance at radio frequencies). C_1, which tunes the aerial circuit, and C_3, which tunes the oscillator circuit, are mechanically connected (ganged) so that the two frequencies change together. C_2 and C_4 are preset variable capacitors used to balance the frequencies when the receiver is being adjusted, or *aligned*.

Figure 15.19 shows a typical IF amplifier stage that might follow the mixer/oscillator of Figure 15.18.

Note that in both these figures the IF transformer is shown surrounded by a dashed box; this indicates that it is constructed as a single component— IFTs are mass produced and are very cheap. The IFT is usually mounted

fig **15**.18 *mixer/oscillator stage, with a tuned collector load*

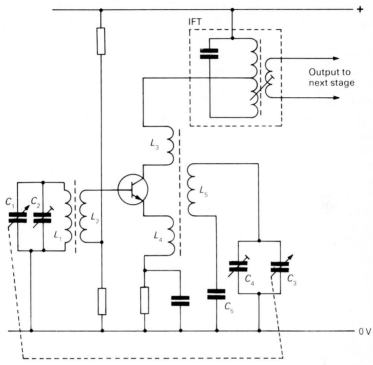

fig **15**.19 *a typical IF amplifier stage*

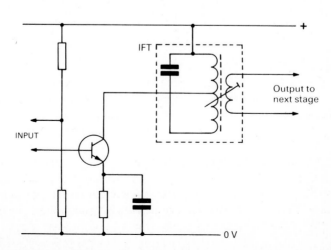

inside an aluminium can which acts as mechanical protection and also as screening against stray electric fields. In practical circuits the aluminium can is connected to supply 0 V, to 'earth' stray signals. The IFT can has a small hole in the top through which the ferrite core can be adjusted to tune the *LC* circuit over a small range—this is used when aligning the receiver. The IFTs are clearly visible in the photograph of the model control receiver (see Figure 15.23, p. 219).

Following two or even three IF amplifier stages is the detector. In AM receivers this is really just a diode, but usually with a slightly more sophisticated smoothing circuit than the single capacitor of the crystal set. A typical detector stage is shown in Figure 15.20. And last, there is an audio amplifier.

fig 15.20 *an IF stage followed by a detector*

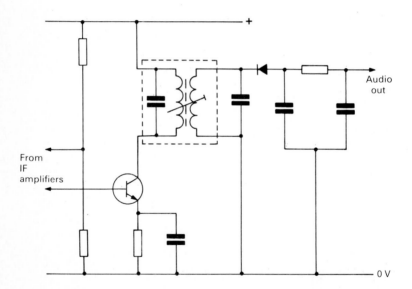

There is one more feature that appears in all but the most rudimentary radio receivers, and that is the provision of *automatic gain control* (AGC). If a sensitive receiver is tuned to a powerful nearby transmitter, the signal may be so strong that it overloads the later IF amplifier stages, causing severe distortion of the sound. To prevent this, AGC is added. The idea— and the circuit—is quite simple, and involves using a proportion of the rectified signal from the diode to control the biasing of the first IF amplifier transistor. A large signal at the detector diode lowers the bias potential at the base of the first IF transistor and reduces the gain—which, of course, reduces the output signal. With the right circuit values the receiver balances out with the correct output and bias levels.

To prevent the AGC responding to transient changes in the audio content of the signal (and increasing the gain during the singer's pauses for breath!), a large-value capacitor is placed across the AGC feedback line to delay the operation of the AGC by acting as a reservoir—the AGC then responds to the average signal level, averaged over a period as long as a couple of seconds.

An implementation of an AGC circuit is shown in Figure 15.21 (refer also to Figures 15.19 and 15.20).

fig **15.21** *application of feedback to provide automatic gain control*

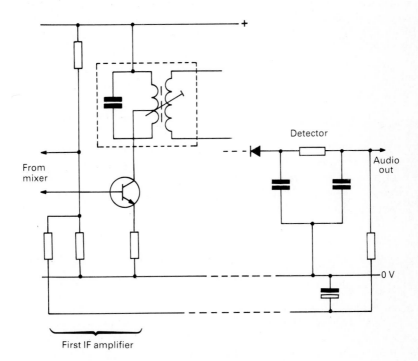

15.4 **A PRACTICAL RADIO-CONTROL RECEIVER**

It is quite possible to construct a working radio control receiver which, when used with the transmitter described at the beginning of this chapter, will have a range of up to half a mile. Commercially-designed radio control receivers generally use specially designed ICs that minimise the number of components required, but this one uses circuits based on discrete transistors. It means that there are a few more parts, but it does have the big advantage that you can see how the circuit works, and find your way through the functioning of every part of the system. The complete circuit of the receiver, from aerial to output, is illustrated in Figure 15.22.

This receiver is a little simpler than a broadcast receiver (and much easier to adjust) because it operates on a fixed frequency. The frequency is set by a crystal-controlled oscillator almost identical to the one used in the transmitter (and to the one in Figure 10.24), apart from a simplified bias circuit using just one resistor. In a broadcast receiver the local oscillator and the tuning circuit are adjusted in frequency together to preserve the correct intermediate frequency, which is then selectively amplified by the IF amplifiers. In the radio control receiver the requirement is for the tuning to be set to a single, very accurately determined frequency, which is why the more expensive crystal-controlled oscillator is used in preference to an LC oscillator.

The crystal is at a lower frequency than the crystal in the transmitter, differing from it by the selected value of the intermediate frequency, in this case 455 kHz. The transmitter crystal was 27.095 MHz, so the required crystal for the local oscillator in the receiver would be:

$$27.095 - 0.455 = 26.640 \text{ MHz}$$

This produces the correct intermediate frequency when mixed with the incoming RF carrier. A signal received from a transmitter working on a slightly different frequency would produce an intermediate frequency higher or lower than 455 kHz, which would not be amplified by the IF amplifiers and would therefore not produce an output.

The incoming radio signal is amplified by a JUGFET. This device is used in preference to a bipolar transistor because of its very high gate resistance, which maximises the efficiency of the aerial tuning circuit (L_2) that helps to reject powerful signals at unwanted frequencies. The output of the oscillator is transformer coupled to the source of the JUGFET (the signal does not need amplification), so that mixing of the radio and oscillator frequencies takes place. Following the mixer there are two IF amplifier stages. Bias for the transistors is obtained entirely from the AGC line – again, slightly simpler than the 'textbook' circuit in Figure 15.21.

The detector stage incorporates a diode between the base and emitter, to rectify the input signal for that stage. The AGC voltage is derived from this stage. Finally, there is a d.c. amplifier fitted with low-pass filters to reject unwanted noise from the earlier stages.

This circuit will work perfectly with the transmitter – but as it stands the transmitter will be sending only the carrier, and there won't be any output! To tune the receiver, a *modulated* 27.095 MHz transmission is needed. A borrowed AM model control transmitter could be used, fitted fig 15.22 *the circuit of a complete superhet receiver; this is a practical design for use with the 27 MHz transmitter described at the beginning of this chapter. The receiver is crystal-controlled and is intended for model radio control work. A suitable decoder circuit is given in Figure 22.23. See also Appendix 1*

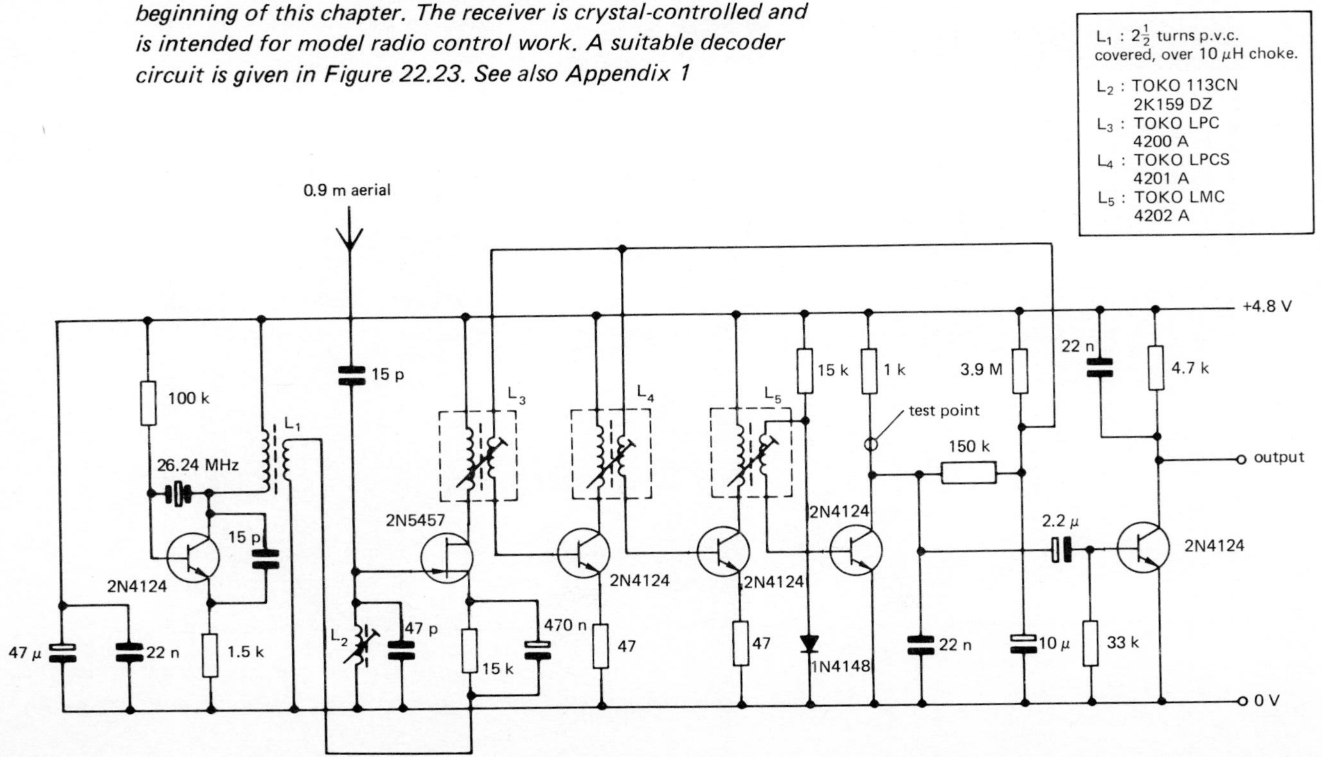

L1 : 2½ turns p.v.c. covered, over 10 μH choke.
L2 : TOKO 113CN 2K159 DZ
L3 : TOKO LPC 4200 A
L4 : TOKO LPCS 4201 A
L5 : TOKO LMC 4202 A
0.9 m aerial
+4.8 V
output
0 V
test point
100 k
26.24 MHz
15 p
2N4124
47 μ
22 n
1.5 k
L1
L2
47 p
2N5457
15 k
470 n
L3
L4
L5
47
1N4148
1 k
3.9 M
150 k
22 n
10 μ
2.2 μ
33 k
4.7 k

fig 15.23   the complete radio control circuit boards – transmitter with
its encoder, and receiver with decoder

with a 27.095 MHz crystal, but the transmitter described in this chapter can be used with a modulation system described in Chapter 22.

To tune the receiver, the 0.9 m aerial must be connected and arranged roughly straight. Connect a voltmeter to the collector of the detector transistor (marked 'test point' in Figure 15.22) via a 4.7 kΩ resistor. Select a range reading around 5 V full scale. When the battery – ideally a 4.8 V nickel-cadmium accumulator – is connected, the meter should show little, in any, deflection. Switch on the transmitter with the aerial collapsed. When the transmitter is moved close to the receiver, the meter will begin to show a deflection. Using a plastic trim-tool, adjust L_2 for the maximum deflection, if necessary moving the transmitter away to keep the meter reading about 2.5 V.

Now repeatedly adjust L_3, L_4, and L_5 in turn to obtain the highest possible meter reading, if necessary moving the transmitter even further away; you may end up with the transmitter several metres from the receiver. When no further improvement can be obtained, the receiver is correctly aligned.

15.5 CONSTRUCTION PROJECT

Sufficient information has been given here (and will be in Chapter 22) to enable you to build the model control transmitter and receiver as a construction project. Although circuit layout is not particularly important, the printed circuit board designs given in Appendix 1 are recommended. The designs are fully compatible with modern radio control practice, and can be used with commercially available servos (see also Figure 15.23).

15.6 TELEVISION RECEIVERS

Having covered the principles of radio transmission and reception, it is now possible to consider the principles and practice of television. Even monochrome (black-and-white) televisions are extremely complicated in the details of their circuits, and television is a complete subject in its own right. In *Mastering Electronics* the systems involved will be explained, but without circuit details, except where this is essential to an understanding of the systems. To include a detailed study of television (and of various other topics, also) would be to make this book at least twice its present size and cost!

15.7 MONOCHROME TELEVISION RECEIVERS

Most people now know that motion picture films produce the illusion of movement by presenting a rapid sequence of still pictures, each one slightly different from the one before it. The eye 'joins up' the pictures and interprets them as a single image in smooth motion, a phenomenon called 'persistence of vision'. Home movies (silent) show 18 frames of film (i.e. 18 pictures) every second. Sound movies shown 24 frames per second.

The movement on the television screen is created in exactly the same way, and the television transmission is sent out as a series of 'still' pictures, at the rate of 25 per second in the UK and many European countries, and at the rate of 30 per second in the USA and elsewhere. (The rate is actually half the frequency of the mains supply in the country in question. In the UK the mains supply is at 50 Hz, and in the USA it is 60 Hz.)

Central to the television system is the TV tube, illustrated in section in Figure 5.8 (p. 65). The bright spot is made to scan the front of the tube continuously, in lines running across the screen, moving down from top to bottom in a pattern called the *raster* (see Figure 15.24).

fig 15.24 *television screen showing the raster*

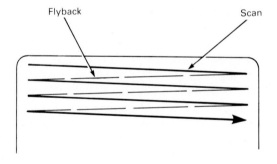

The brightness of the spot is reduced to nothing during the right-to-left 'return' stroke, the *flyback*. In the UK the screen has 625 horizontal scan lines in the raster; it is slightly different in some other countries. Unfortunately, a scan of this sort would produce flicker, even at 25 frames per second, whereas a motion picture film running at the same speed would not. The reason lies in the way the television picture is drawn from the top downwards, whereas the motion picture projector uncovers the whole of each frame virtually at once. To reduce the flicker, which has an odd appearance of running from the top down the screen, *interlaced scanning* is used, and the picture is produced in two halves, each taking 1/100 th of a second (UK). The electron beam scans the screen as above, but only with $312\frac{1}{2}$ lines. When it gets to the middle of the bottom of the screen (the $312\frac{1}{2}$th line!) it returns to the middle of the top and scans the picture again, but *between the first set of lines*, 'interlacing' the second scan with the first. Using this technique makes the amount of flicker hardly noticeable. Interlaced scanning is illustrated in Figure 15.25, though with an 11-line screen instead of 625, to make the diagram clearer.

fig 15.25 *interlaced scanning*

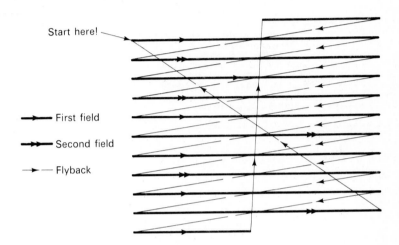

It is important to realise that in a monochrome television the phosphor on the screen is completely homogeneous; the lines are merely a product of the scanning that produces the raster.

Two time periods are critical to the television: the time the beam takes to scan the width of the picture—the line frequency; and the time taken for one frame—the frame frequency. Two oscillators within the receiver generate these frequencies, and they are called, respectively, the *line timebase* and the *field timebase*. The line timebase has to scan the screen $25 \times 625 = 15\,625$ times every second, so has an operating frequency of 15.625 kHz, and the field timebase scans the picture 50 times every second (twice for each frame; remember the interlaced scanning), so operates at 50 Hz (the mains frequency—you will see why this is a good idea later).

The timebases have not got to operate at about the right frequency, they have to work at *exactly* the right frequency, to scan the screen in exact synchronism with the cameras at the studio. For as the electron beam sweeps across the screen it is modulated so that the brightness of the screen is changed continuously to produce a picture. Figure 15.26 shows how a picture is built up.

Clearly the timebases have *got* to be exactly in synchronism with the timebases of the cameras at the broadcast studio, or the TV picture will be chaotic, with no recognisable images at all. The transmitted signal

fig **15.26** *the brightness of the spot forming the raster is modulated to provide pictures on the television screen*

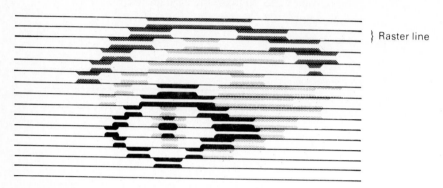

} Raster line

therefore includes information about the timing of the line and field scanning, and circuits in the receiver extract this information and use it to *synchronise* the two timebases.

Every single sweep of the line timebase is *triggered* by a pulse in the signal received by the television. An RF carrier from the transmitter is modulated with a waveform corresponding to the brightness of the line of the TV picture. Figure 15.27 shows part of this waveform, corresponding to two lines of the picture.

fig **15.27** *the waveform of a television picture broadcast; this is the waveform that is recovered from the AM signal received by the television*

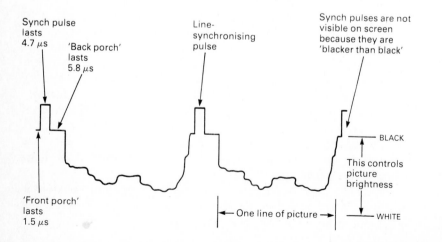

Synch pulse
lasts
4.7 μs

'Back porch'
lasts
5.8 μs

Line-
synchronising
pulse

Synch pulses are not
visible on screen
because they are
'blacker than black'

BLACK

This controls
picture
brightness

'Front porch'
lasts
1.5 μs

One line of picture

WHITE

Compare the waveform in Figure 15.27 with the one in Figure 15.28, which shows two lines of a picture consisting of four vertical stripes, progressively darker from left to right.

fig **15.28** *two lines of a television picture showing four vertical bars of increasing density*

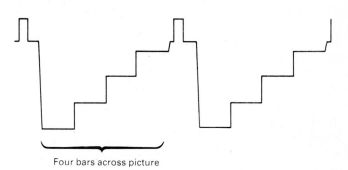

Four bars across picture

The *line-synchronising pulses* each trigger the line oscillator into a single sweep across the picture, or more commonly are used to synchronise an oscillator to run at a precise speed. Although in theory it would work to have an oscillator that is triggered for each line, in practice this would mean that any tiny interruption in the received signal would result in the loss of a synch pulse, and the loss of overall synchronisation for the rest of that field.

At the end of each field there is a special series of line-synchronising pulses that also trigger the *field synchronisation*. The actual sequence is quite complicated and is shown in Figure 15.29.

fig **15.29** *the synchronising pulse sequence at the end of each field*

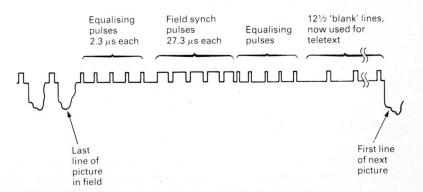

Equalising pulses 2.3 µs each

Field synch pulses 27.3 µs each

Equalising pulses

12½ 'blank' lines, now used for teletext

Last line of picture in field

First line of next picture

Following the last line of the picture in any particular field there are five *equalising pulses*, followed by five *field-synchronising pulses*, followed by another five equalising pulses. The exact function of the equalising pulses is subtle: suffice to say here that it makes the circuits in the receiver simpler. The five field-synchronising pulses are detected by the receiver circuits—they are five times longer than 'ordinary' synch pulses—and trigger the next sweep of the field timebase. Once again, a synchronised free-running timebase is used to prevent the picture collapsing entirely if the signal is lost momentarily.

Following the pulse sequence for field synchronising there are a further $12\frac{1}{2}$ 'blank' lines that are not used for the picture. These are put in to give the electron beam in the tube time to fly back to the beginning of the next field at the top of the picture. When all these various pulses and blank lines are taken into account, a total of twenty lines are 'lost' for each field, i.e. forty for each frame. For a television system like that used in the UK, having 625 lines per frame, only 585 lines actually appear on the screen.

A system diagram for the line and field scan of a television receiver is given in Figure 15.30.

fig **15.30** *system diagram of the synchronisation and scanning circuits of a monochrome television receiver*

15.8 **TELEVISION SOUND**

The signal for TV sound is transmitted on a completely separate carrier; in the UK it is 6 MHz higher in frequency than the carrier used for the vision. The two carriers are amplified together in the receiver IF amplifier circuits, and then combined in a mixer. Since the sound is frequency-modulated and the vision is amplitude-modulated, the result is a frequency-modulated difference frequency, centred at 6 MHz, and carrying the sound. This then goes its own way, being amplified separately, detected with an ordinary FM discriminator (phase-locked loop, etc.; see above), fed to an audio amplifier and finally to the speaker.

15.9 TELETEXT

The information used to update teletext systems in receivers equipped to show 'Ceefax' and 'Oracle' (in the UK) is transmitted in the form of binary code, a series of pulses, much more closely spaced than the line-synch pulses. Teletext arrived after the television standards were all established, and made use of a feature included for other reasons—the blank lines following the field pulse sequence. The 'blank' lines are no longer blank, but are filled with the coded binary information. The teletext system selects the binary pulses and feeds them into a digital system that processes and displays the teletext information on the screen when required.

Only part of the teletext information is transmitted at any time, for it would be impossible to fit in the huge amount of binary code, even over $12\frac{1}{2}$ lines. The teletext system therefore includes a computer memory circuit that selects and holds the required information. Pressing the keys for a new teletext page makes the system 'look at' the pulses, each group of which begins with a code number (in binary) corresponding to a page of text. When the teletext system detects the number you have selected, which might take several seconds, it feeds the binary information into its memory, changing the picture on the screen. The picture then remains in the memory—and on the screen, if you want it—but is updated by the teletext system every time the selected number is detected in the binary coded information.

Complex circuits are used to mix teletext and vision, display different things on different parts of the screen, and to provide a 'newsflash' facility that puts information on the screen only when a change in content for a particular page is detected.

In theory, the amount of information that can be carried via teletext is unlimited, but the more different pages that are transmitted, the longer the delay in selecting a new page, as information for that page will come up less frequently.

15.10 THE COMPLETE TV RECEIVER

A system diagram for a complete TV receiver is shown in Figure 15.31.
The only circuit we shall consider, even briefly, is the *line output stage*.

fig **15.31** *system diagram of a monochrome television receiver*

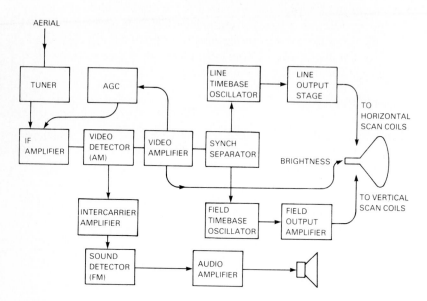

This is the oscillator that provides the waveform required to make the electron beam in the tube scan the lines from one side of the tube to the other. As can be seen in Figure 15.30, the waveform is a 'sawtooth'; the waveform is applied to the horizontal scan coils, and relatively high voltages are needed. The flyback has to be *fast*, and the voltage changes very abruptly. It is possible to make good use of this. In order to produce the high voltage necessary to drive the horizontal scan coils, a transformer is used, driven by the line oscillator. An extra secondary winding is put on the transformer, called the *line output transformer* (LOPT), and with a large number of turns this winding can produce a very high voltage indeed, enough, in fact, to supply the final anode of the picture tube (see Figure 5.8 on p. 65). The line output transformer is therefore a vital—and dangerous—part of the power-supply circuits, as well as part of the line-scanning system.

15.11 COLOUR TELEVISION RECEIVERS

The problems facing the designers of the colour television system were formidable. The television system was well established with monochrome receivers, and it was important that the colour television system used would be cross-compatible. That is, a black-and-white receiver should be capable of receiving a colour transmission and would reproduce it correctly (in black and white); and at the same time, a colour receiver should reproduce a black-and-white transmission as well as a monochrome receiver.

The extra information required for the colour receiver therefore had to be 'fitted in' round the existing signal, and in such a way that it would not interfere with the operation of a monochrome receiver. The colour signal also had to be squeezed into the available bandwidth, which had been established as 8 MHz.

Since the eye is far less sensitive to colour than it is to brightness, it proved possible to transmit the colour information over a rather restricted bandwidth, so that the *colour* on a colour television is actually rather fuzzy. But if the brightness is controlled with a relatively wide bandwidth signal, giving a sharp picture in terms of brightness, the resulting picture is perfectly acceptable.

Figure 15.32 shows a waveform corresponding to one line of the same four vertical stripes illustrated in Figure 15.28 above, but with colour information added. The colour signal is actually coded during the vision information period, and at first sight seems to be likely to interfere with a monochrome receiver. But remember that the frequency of the colour signal—the *chrominance* waveform—is high, at 4.43 MHz, and would not

fig 15.32   *the waveform of one line of a colour television picture (the waveform is for a picture consisting of four vertical bars, like the one in Figure 15.28)*

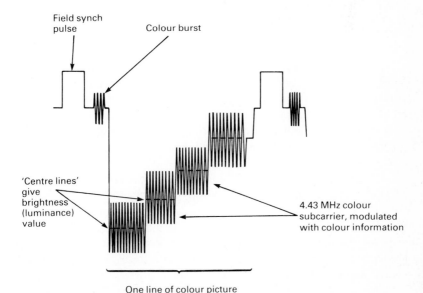

One line of colour picture

be resolved by the circuits of the monochrome receiver, which would *average* the signal. The average is shown dotted in Figure 15.32, and is obviously the same as the brightness waveform—called the *luminance* waveform—of the bars in the illustration in Figure 15.28.

By means of a theoretically (and practically) complex method known as *quadrature modulation*, information about three colours can be obtained from the frequency-modulated chrominance *subcarrier*. (The waveform is called a 'subcarrier' because it is a carrier waveform, derived from modulation of another carrier of higher frequency.) The complete range of colours can be reproduced from these three colour signals, as we shall see. It is, however, necessary to have a 4.43 MHz reference oscillation, not only at precisely the same frequency as the reference oscillator back at the transmitter, but also exactly in phase with it (see Figure 15.12 on p. 206 for a reminder about phase relationship). This technical miracle is achieved by means of a crystal oscillator in the receiver, synchronised in frequency and phase with a short (ten cycles) burst of the 4.43 MHz transmitted just after each line-synchronisation pulse. The oscillator can be relied on to remain at the right frequency and phase for one line duration, 64 μs! The short burst of reference frequency is called the *colour burst*.

Even with a system like this, slight changes in the transmitted waveform, caused by weather conditions, reflections from aircraft, etc., can cause phase errors between the colour burst and the actual picture colours. The effect of this is a drift of the whole colour spectrum towards red or blue—not desperate, but enough to notice. Despite this problem, this system was adopted in the USA, the first country to have a broadcast colour TV network. The system is called the NTSC system, after the American National Television Standards Committee. (Opponents of the system claim that NTSC actually stands for 'Never Twice the Same Colour', but this is not true.)

The system adopted in the UK and many other countries, with the benefit of learning from the pioneers' mistakes, is known as PAL, which stands for *Phase Alternation by Line*. This overcomes the colour shift of NTSC by the ingenious method of turning round the modulation system on alternate lines. Colour shifts still occur, but alternate lines are changed towards blue and then red, and blue and then red, etc. The eye merges the colours and the overall impression is correct.

Very large colour shifts still look odd. The effect is a bit like a multi-coloured venetian blind, so a modification of the PAL system, known as *PAL-D* was introduced. PAL-D actually adds the colour signals of alternate lines electronically, by storing a whole line in a device called a *delay line* while waiting for the next line. The colour on the screen is always the product of two alternately coded signals, and the result is effectively perfect colour stability.

15.12 COMBINING COLOURS

The information recovered from the transmitted signal is about three colours only, but it is possible to use the three colours to make every colour in the visible range. In just the same way that the artist can make every colour (if he wants to) by mixing primary colours, so the three primary colours of the colour TV are mixed in different proportions to produce the whole range of colours. The three colours used are *red, green* and *blue.* Mixed together in equal proportions, these make white (white light is 'all the colours of the rainbow' mixed together), and other colours can be made by mixing two or three in different proportions. For example, green and red mixed together produce a bright yellow. This process is called *additive mixing*, and is not the same as the mixing of artist's colours, which is called *subtractive mixing.* The artist uses pigments that reflect only certain colours from the white light falling on them—they subtract colours. In the TV tube, colours are mixed—they are added together.

There are three factors that affect the picture on the screen. These are *brightness, hue* and *saturation.* Brightness is straightforward, and on the screen is controlled by the increase or decrease of all three colours simultaneously. Hue (colour) is controlled by the balance of the three colours, as described above. Saturation refers to the 'strength' of the colour, the amount of white light that is added to the basic colour. Red, for example, is a saturated colour. Pink is red mixed with white, and could be said to be a less saturated red. White light is not of course produced by the TV receiver, so saturation is in fact controlled by the difference between the brightness of the three primary colours; it is convenient to think of it as the colour mixed with white, however.

15.13 THE PICTURE TUBE

The colour TV picture tube is a development of the monochrome tube illustrated in Figure 5.8 (p. 65). The overall shape is the same, but there are three separate electron guns. In the most popular type of tube, the slot-mask tube, the three electron guns are arranged in a row, horizontally. The guns produce three electron beams, one for each colour, and although the beams are all deflected together by the vertical and horizontal scan coils, the brightness of the beams can be controlled separately.

All three beams therefore scan the front of the tube together, to produce a raster. Obviously, it is not possible to have coloured electron beams, so the colour is produced in phosphors on the front of the screen. Phosphors can be made almost any colour, and the right shades of red, blue and green can be produced easily, if expensively. The raw materials

fig 15.33 *principle of the slot-mask colour television tube*

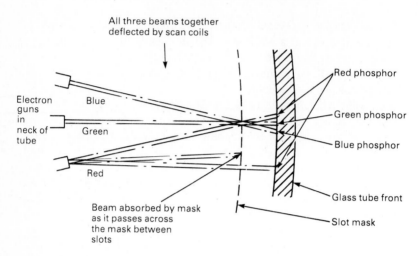

All three beams together
deflected by scan coils

Electron
guns
in
neck of
tube

Blue

Green

Red

Red phosphor

Green phosphor

Blue phosphor

Glass tube front

Slot mask

Beam absorbed by mask
as it passes across
the mask between
slots

(*Not to scale!*)

for the coloured phosphors actually contribute significantly to the price of a colour TV tube.

The trick is getting the 'red' beam to affect only the red phosphor, the 'blue' beam to affect only the blue phosphor, and the 'green' beam to affect only the green phosphor. This is done by simple geometry. Figure 15.33 shows the system diagrammatically, viewed from the top of the tube.

The slot-mask is fixed firmly in place behind the tube, and the relationship of the positions of the slots and of the phosphor strips on the back of the tube is such that the 'red' electron beam can fall only on the red phosphor, etc. Figure 15.34 is a perspective sketch of the tube, from which you can see that the slots are quite short, and are 'staggered' to produce an interlocking pattern. This arrangement is used because it is physically strong—great rigidity of the mask is clearly important—and because it gives a reasonably large ratio of 'slots' to 'mask'. The more transparent the mask is in this respect, the better, for electrons which hit the mask rather than go through the slots are just wasted power, and serve only to heat the mask. The larger the slots can be made in relation to the mask, the brighter the picture will be.

The slot-mask principle is used in the popular *precision in-line* (PIL) tube, which is manufactured complete with scan coils as a single unit. This, together with extremely sophisticated scan-coil design, produced in the first place with computer-aided design techniques, has led to a tube that is very simple to use, all the most critical alignments having been built in at the manufacturing stage.

fig 15.34 *perspective drawing of a slot-mask colour TV tube*

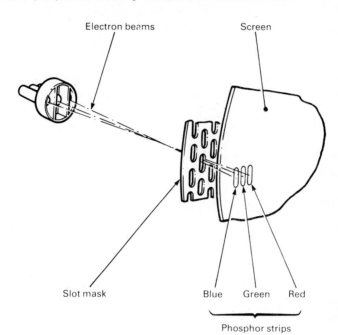

Electron beams

Screen

Slot mask

Blue Green Red

Phosphor strips

(Not to scale!)

A variation of the slot-mask tube is the Sony Trinitron® tube. This uses an *aperture-grill* instead of a slot-mask. The aperture-grill consists of vertical strips, like taut wires, with complete vertical phosphor stripes in different colours printed on the back of the tube. The Trinitron® tube has very good mask transparency and gives a good picture brightness. The Trinitron® tube also includes an improvement to the electron gun system, in which the three electron beams all pass through a single common point in the neck of the tube, so that they all appear to originate from the same point. This makes adjustments of the scanning system easy, and improves the focus (the sharpness of the picture on the screen).

Earlier colour TV receivers used 'delta-gun' tubes, which had the three electron guns arranged in a triangle, and circular holes in the shadow-mask tube. These tubes are not now used in new receivers, as they are rather difficult to adjust and give poorer picture brightness for a given beam power than either the PIL tube or the Trinitron®.

There is not space here to give more than the briefest glance at colour tele- vison receivers. The circuitry is, in both theory and in practice, very complicated, and extensive use is made of special-purpose integrated circuits to reduce the 'component count' and to make the receivers easy and cheap to manufacture. In real terms a colour television costs less that its monochrome counterpart of thirty years ago, and gives incomparably better results.

Television servicing, especially colour television servicing, is highly specialised, and should never be attempted by the inexperienced or untrained. The voltage on the final anode of a colour receiver will be more than 25 kV, and is usually lethal.

Before leaving the subject of colour TV, get hold of a pocket magnifying glass, and take a close look at the surface of a colour TV tube, first with the receiver turned off, then with it turned on.

QUESTIONS

1. What is meant by wavelength?

2. What kind of oscillator is used for radio control applications? Why?

3. Why are germanium diodes, as opposed to silicon diodes, often used in the detector stage of radio receivers?

4. Describe amplitude modulation, and compare it with frequency modulation.

5. Draw a circuit for a 'crystal set' radio receiver, and describe how it works.

6. What are advantages of the superheterodyne receiver?

7. Why do radio receivers and transmitters usually have to be 'trimmed', or adjusted, before they can be used?

8. Why do television systems use interlaced scanning?

9. Describe the functions of the television line and field synchronising circuits.

10. Which part of the television's circuits generates the EHT supply for the final anode of the cathode ray tube?

11. Which colours are used to make the whole range of visible colours in a colour television tube?

12. Sketch the internal layout of a colour TV tube, showing how the electron gun assemblies produce the relevant colour on the front of the tube.

OPTOELECTRONICS

Although devices that combine light with electronics are by no means new, the name 'optoelectronics' is recently coined. The same revolution that replaced the valve with the transistor, and subsequently complex transistor circuits with integrated circuits, produced an increase in the number and types of devices combining light and electronics. The term 'optoelectronic' came into use as a description for what has now become a large and important branch of electronics technology.

In a way the television picture tube and the neon lamp are both optoelectronic devices, but the name does not apply to them in general usage because they are 'vacuum-state', not solid-state, devices. Optoelectronics refers to the new generation of solid-state components, and also to some of the components used with them.

16.1 LIGHT-EMITTING DIODES

Easily the most common optoelectronic device is the light-emitting diode (LED) which was discussed briefly at the end of Chapter 7. The LED is cheap (two or three for the price of a cup of coffee) and uses little power, so has found applications wherever an indicator lamp is required. Electrical and electronic equipment of all kinds use LEDs as indicators. The most usual colour is red, but yellow and green are also available. Like the *pn* diode of the normal non-light-emitting variety, the LED exhibits a forward voltage drop when conducting. This varies according to the type of semiconductor used, so voltage drops differ slightly. Typical values are around 2 V for red and 2.5 V for green or yellow. The usual type of LED used as an indicator light is 0.2 inch diameter or $\frac{1}{8}$ inch diameter. The light output is usually around 2 mod for a current of 10 mA: not actually very bright, but enough for an indicator in all but the very brightest ambient light.

There is a photograph showing typical LEDs in Figure 7.13 (p. 92). LEDs are very easy to use in circuit, and have an indefinite life if used correctly. There are just a few rules that need to be remembered when using them:

1. The LED has no inherent current limitation, and must have a resistor (or something) in series with it to limit the current passing through it to the manufacturer's recommended safe value.

2. LEDs have a low reverse breakdown voltage, so must be protected from voltages applied in the 'wrong' direction. The simplest way to do this is to connect a diode in parallel with the LED, but facing the other way. Figure 16.1 shows a typical circuit that permits the LED to be operated with an alternating current supply. The value of the resistor is chosen to provide an average current suitable for the LED. At mains frequency this circuit will cause a perceptible flicker, as the LED is being pulsed fifty times a second, with an 'off' period equal to the 'on' period.

3. The LED emits light over a fairly narrow band of wavelengths, so it is unfortunately not practicable to use filters over the LED to produce different colours. Red, yellow and lime-green seem to be the extent of the repertoire so far.

4. The LED has no time lag on the production of light (such as there is on a normal tungsten filament lamp) and so can be switched on and off very rapidly.

5. A small LED can generate a very bright pulse of light if it is brief. The standard 0.2 inch LED can be used to produce short pulses at much higher operating current than the rated maximum for continuous current. The manufacturers will give information, but typical figures for a 0.2 inch red LED would be 300 pulses per second, 1 μs in duration, with pulse current of 1 A.

fig 16.1 *circuit for operating a LED from an a.c. supply*

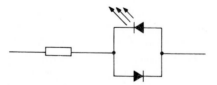

The LED can easily be combined with a small integrated circuit; *flashing LEDs* use an internal IC, as do *constant-current LEDs*, which require no current limiting resistor and work over a range of voltages.

LEDs are often used in multiple displays – a LED 10-bar DIL array is shown in the photograph in Figure 7.13 (p. 92). The circuit for this is shown in Figure 16.2.

fig **16.2** *a 10-LED bar graph display*

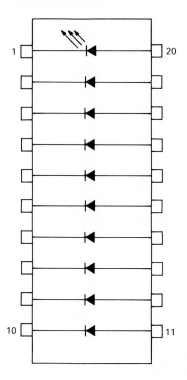

The bar array is designed so that it can be *stacked* with other identical arrays, to make a *bar graph* display of any length required. Such indicators are often used for record level meters on hi-fi tape decks, as they are cheaper, more accurate, more robust and have more 'sales appeal' than ordinary mechanical meters. Special-purpose ICs are made for driving bar arrays.

LEDs are also used for the familiar *7-segment display*, of the kind commonly used in digital alarm clocks. The 7-segment display can be used to produce any of the digits 0–9, and a very limited number of letters. The basic form of the display, together with a set of digits and letters, is given in Figure 16.3.

A typical alarm-clock display, equipped with hours, minutes and seconds, would require six 7-segment displays. If they were all connected like the one in Figure 16.3, with seven connections to the bars and a common anode or cathode, there would be eight connections to each digit, or forty-eight connecting wires altogether. This is not very economical, and is wasteful of power, when you consider that at 12 : 58 : 58 there will be no less than thirty-one segments illuminated, each taking about 15 mA,

fig **16.3** *a basic 7-segment display, along with some of the characters that can be produced*

a total current consumption approaching half an amp, and a power dissipation, in the LEDs alone, of nearly 1 W. In practice the display is *multiplexed*, and is connected as shown in Figure 16.4.

This cuts the number of wires down to thirteen, and only adds one more wire for each extra digit. The circuit driving the display is specially designed to make use of the display economically, and in practice only one digit is illuminated at any one time. The cathodes of the relevant segments for the first digit are switched to the power supply, and the digit-1 anode is made positive for a short time. The digit-1 anode is then disconnected and the cathodes switched in a suitable pattern for the second digit. The digit-2 anode is now made positive for a short time, and the sequence repeated over and over for each digit in turn.

fig **16.4** *multiplexing a 6-digit 7-segment display*

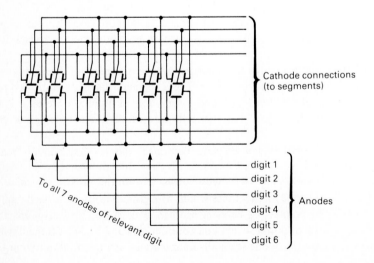

Only one digit is ever on at any one time, but if the sequence is repeated at a high-enough speed, the human eye's persistence of vision will make it look as if the six digits are all lit up together. The multiplexing frequency is usually in the order of 1 kHz or so.

LEDs are commonly produced to emit light in the infra-red region, and infra-red LEDs are actually more efficient in output for a given current than LEDs operating in the visible part of the spectrum. Infrared LEDs are widely used for remote control and sensing applications, described later in this chapter.

16.2 LIQUID CRYSTAL DISPLAYS

Although LEDs are relatively efficient when compared with filament lamps they nevertheless use a lot of power compared with some types of IC. Early digital watches used LED displays, and even with low-brightness, high-efficiency multiplexed displays, the LED display still used a few milliamps, much too much to be left on all the time. These early watches therefore featured a 'black face', on which the digital time was displayed for a second or so when a button was pressed. Manufacturers of watches (and calculators) searched for a *really* low-power display system that could operate for a long time—at least a year—from a battery small enough to go inside a wristwatch. The solution proved to be the *liquid crystal display* (LCD).

The LCD is unlike the LED in that it does not emit light, but is read, like the page of this book, by reflected light. The typical form is similar to that of a 7-segment LED display, and is illustrated in Figure 16.5.

fig 16.5 *a typical 7-segment LCD*

The connections to the LCD are reminiscent of the LED display, but the actual operation is completely different. The usual kind of LCD, called a 'twisted nematic' display, works by means of polarising light. (There is another kind of LCD, called a 'dynamic scattering' display, but it is seldom used as it tends to have a shorter life, and produces rather curious milky-white characters against a dark background—harder to read than the more common twisted nematic display.) Most people are familiar with polarising sunglasses. These polarise the light passing through the lenses so that the light waves move in one plane only. If an identical pair of sunglasses is placed in front of the first, there will be only moderate darkening of the beam passing through both, because the planes of polarisation are the same. If, however, the second polarising lens is placed so that the plane of polarisation is at right angles to the first one, the light ray will be blocked completely. Figure 16.6 shows the principle.

fig 16.6 *if light is passed through a polarised filter, it will then pass through subsequent polarising filters only if the plane of polarisation is the same; if the planes of polarisation are at right angles, the light will be blocked*

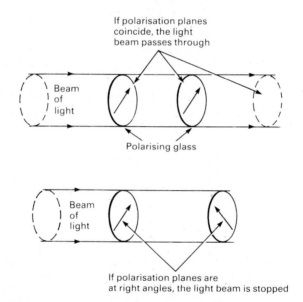

A cross-section of the LCD is shown in Figure 16.7a. Two thin pieces of glass are separated from each other by a plastic seal, and are held a precise distance apart, in the order of $10\,\mu m$ or so. The space between them is filled with the liquid crystal (*really* a liquid). On the inside of each piece of glass is a thin coating of a transparent conducting film. The pattern in which the film is printed produces the required display (see Figure 16.7b).

The liquid crystal has the property of rotating the plane of polarisation of light passing through it. The amount of rotation depends on the precise composition of the crystal and on the thickness of the crystal—$40\,\mu m$ will

fig 16.7

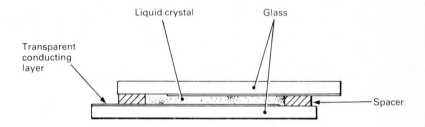

Liquid crystal   Glass

Transparent
conducting
layer

Spacer

(a) *a sectional view of an LCD*

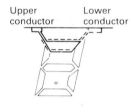

Upper   Lower
conductor   conductor

(b) *arrangement of transparent conductors to operate one
segment of the display*

give a full $360°C$ rotation in a typical material. Thus $10\,\mu$m thick crystal
will rotate the plane of polarisation through $90°$.

Behind the LCD is placed a piece of reflective polarising material (like
polarised sunglasses with a mirror behind them!), and in front a piece of
transparent polarising material. The planes of polarisation of the two
polarisers are at right angles to each other, and in normal circumstances
any light reflected off the rear polariser would be blocked by the front
one. However, the $90°$ twist imparted by the liquid crystal rotates the
plane of polarisation of the light so that most of the light falling on the
display from the front is reflected back again, and the LCD appears to be
completely transparent. You can even see the polarising reflector behind
the display if you look carefully—but try looking at the display of a LCD
watch through polarising sunglasses, rotating the glasses to see the effect
of changing the plane of polarisation!

If the display is energised by means of an electric voltage applied
between the two transparent conductors controlling one of the segments,
the situation changes. The liquid crystal is an insulator, so there is practically no current flow; but the electric field that appears between conductors that are opposite to each other affects the crystal, changing its regular

structure in such a way that it no longer rotates the plane of polarisation of light passing through it. Light reflected off the rear polariser therefore reaches the front polariser with its plane of polarisation 90° out—and fails to pass through it. Viewed from the front, the energised segments appear black against the bright reflected light of the polariser. The principle of operation is shown in Figure 16.8.

The LCD has the big advantage that, like the field-effect transistor, it does not need a current flow to work, and power requirements are there-

fig **16.8** *principle of the LCD (twisted nematic type)*

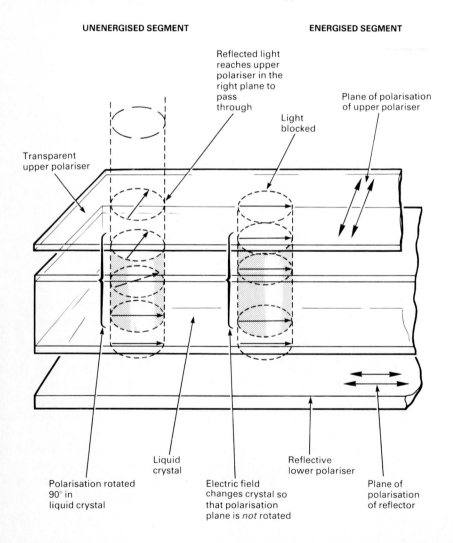

UNENERGISED SEGMENT   ENERGISED SEGMENT

Reflected light reaches upper polariser in the right plane to pass through

Plane of polarisation of upper polariser

Light blocked

Transparent upper polariser

Liquid crystal

Reflective lower polariser

Polarisation rotated 90° in liquid crystal

Electric field changes crystal so that polarisation plane is *not* rotated

Plane of polarisation of reflector

fore tiny. A large (12 mm high) LCD with four digits would draw less than 5 μA from a 3 V supply.

In addition, the LCD works in bright light (the LED display watch is hard to read in the sunshine) and can be illuminated with a yellow LED or tiny incandescent lamp for the rare occasions in which it will need to be read in the dark.

Good ideas are rarely without drawbacks, and the drawbacks of the LCD are relatively few. But it does require an *alternating* voltage to energise the segments, not because they will not work with direct voltages, but because a direct voltage damages the crystal over a relatively short time. Where the LCD is driven by an IC designed for the purpose, this is no problem at all, but in circuits made with discrete components it calls for slightly more complication. Second, the LCD is not as robust as the LED display, and a hard knock will break the glass. In watches, careful case design minimises the chance of breakage.

Since the shapes of the energised segments of the LCD are printed on to the glasses, there is no limit to the kind of picture that can be built up. Where letters and number are required, it is practical to use a *dot matrix* LCD. The details vary, but the design shown in Figure 16.9 is widely used. A wide range of numbers, letters and special characters can be built on to the dot matrix.

fig **16.9** *dot matrix displays*

photograph by courtesy of Hewlett-Packard

The same limitations on connections apply to LCDs that apply to LED displays, although LCDs are less commonly multiplexed than LED displays. Multiplexed displays are made only where very large numbers of separate segments or dots are needed, as the LCD uses so little power that multiplexing is unnecessary, except to save connections.

LCDs are quite unlike LED displays in the speed of response, for the LCD is a relatively slow device, taking at least tens and sometimes hundreds of milliseconds to respond.

16.3 PHOTO-SENSITIVE DEVICES

The photo-conductive cell (photoresistor)

All semiconductors are sensitive to light, to a greater or lesser extent. It is for this reason that transistors and diodes are packaged in light-proof encapsulations, either metal or plastic.

Certain components make use of the light sensitivity, and are designed to be used as light-detecting devices. There are several different basic devices, but the simplest and one of the most commonly used is the photo-conductive cell, or *photoresistor*.

When light falls on a semiconducting material, some of it is absorbed. The energy that this imparts produces an electron–hole pair, and the free electron and hole are available for carrying current through the semi-conductor, reducing the electrical resistance. The electron and hole are usually formed in the valence band (see Chapter 6), which is where the current flows. The amount of energy delivered by the light falling on the semiconductor is important, and since light of different wavelengths provides different amounts of energy, the semiconductors are selective in the frequencies of light that they respond to. Almost all semiconductors, including germanium and silicon, respond to light in the infra-red region. This is generally unimportant, except where the response of the photo-resistor must parallel the light response of the human eye.

Just one material responds in the region of visible light, and this is *cadmium sulphide*. Cadmium sulphide (usually referred to by its chemical symbol of CdS) is sensitive to about the same range of frequencies as the eye, and is commonly used in camera exposure control systems.

Physically, the CdS photoresistor is no more than a thin layer of cadmium sulphide on a ceramic base, with metal (usually aluminium) connections printed on top. The electrodes are made with a characteristic interlocking comb layout, to maximise the length of the 'junction' in relation to its width. Figure 16.10 illustrates the design.

In the dark the CdS photoresistor has a high resistance, and the ORP12, the standard 'large' photoresistor, has a *dark resistance* of $10\,\text{M}\Omega$. In bright sunshine the resistance can drop as low as $150\,\Omega$, or even less. In

fig **16.10** *a typical photoresistor (CdS cell)*

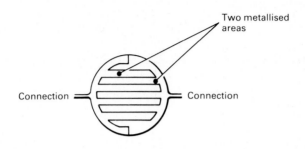

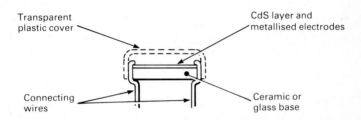

terms of electronics this is a massive range, which probably accounts for the rather small number of different designs produced by the different manufacturers. The ORP12 and the smaller ORP60 take care of most requirements.

CdS photoresistors can dissipate moderate amounts of power, for example the ORP12 can dissipate up to 200 mW. They can also be used with relatively high voltages; the ORP12 can survive 110 volts maximum. Because the design is symmetrical, the photoresistor is unaffected by polarity of applied voltage and can be used with alternating or direct voltage supplies.

Figure 16.11 illustrates the CdS photoresistor in a typical application, turning a mains voltage lamp on when it gets dark and off again at dawn. The circuit symbol is shown with the three arrows denoting 'light'.

The circuit of Figure 16.11 is completely practical and can be made as a project. Notice, however, that the circuit is *live to the mains supply* and *must be properly insulated*. If in any doubt about safety, do not construct this circuit, or at least get expert advice about insulation. Note also that the 7 W resistor gets *hot*. This circuit uses a thyristor to control the mains; the triac is described in Chapter 17. The light-control circuit is given as an example of a typical 'commercial' application. It is interesting to

fig **16.11**   *a practical circuit to turn a mains-operated light on at dusk and off at dawn: caution—see text before attempting to build this project*

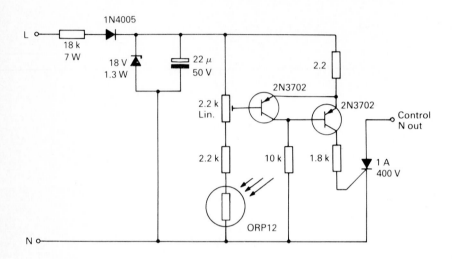

compare the circuit of Figure 16.11 with that of Figure 16.12, which does a similar thing, though rather more safely.

In the circuit of Figure 16.12 a transformer power supply operates the same basic phototransistor circuit, and the mains is switched by a relay (also dealt with in Chapter 17). The photoresistor circuit is operated at low voltage, and less heat is dissipated. For reasons of cost, the circuit of Figure 16.11 would be preferred by a manufacturer, who could make sure that the insulation met the required standards. Figure 16.12 makes a good project; even so, you must observe stringent safety precautions as mains is still present on the primary of the transformer and also the relay.

If you are in *any way* unsure about safety, then the part of the circuit in the dashed box in Figure 16.12 can be made, and will work with a 12 V battery.

Although photoresistors have many uses, they respond to changes in light rather slowly, and so are unsuitable for control applications, or other applications where rapid response to light level is required.

Photodiodes
A photodiode is structurally very similar to a normal *pn* junction diode, though there may be a mechanical differences, brought about by the necessity to maximise the area of the junction that can be exposed to light. Photodiodes are used in a reverse-biased mode, and the leakage current will

fig 16.12 *a lighting-control circuit using a transformer and relay for added safety*

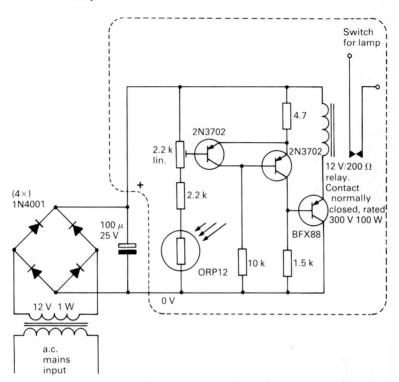

then depend on the amount of light falling on the device. Photodiodes are useful for measurement applications, since the leakage current is directly proportional to the light intensity over a wide range.

Silicon is generally used for photodiodes, so the peak response to light is in the infra-red region. The actual amount of current is also rather small. A typical photodiode might have a dark current of 1.5 nA, and an output current in bright sunshine of 3.5 μA. Quite substantial amplification is therefore required for most applications, and this would usually be provided by an op-amp.

Phototransistors

In the same way that the photodiode is very like a normal *pn* junction diode, so the phototransistor is very like a normal bipolar junction transistor. Apart from a transparent encapsulation (a metal case with a glass or plastic window in the top, or, more cheaply, solid transparent plastic) and possibly mechanical changes to expose more of the junction area to light, there is no difference. The base–emitter junction is either left discon-nected or is slightly reverse-biased, and the junction operates as a *pn* photodiode. However, the collector current is amplified in the normal way, and so may be up to two or three hundred times larger than the output current of the photodiode. A typical silicon phototransistor might have a leakage current (base open-circuited) of a few tens of nanoamps, and a collector current of $500\,\mu A$ in bright sunlight.

It is also quite common to combine two transistors in one, to make a *photo-Darlington*, such as the 2N5777 illustrated in Figure 16.13. In the Darlington pair configuration the gain of the two transistors is multiplied together, so the output of the photodiode is multiplied by at least 10 000. Photo-Darlington transistors are inexpensive and are usually preferred to a photodiode or phototransistor with extra amplification. The photo-Darlington can be connected directly to CMOS logic circuits when used with reasonable illumination levels—this is useful for applications involv-ing counting objects, as you will see in Chapter 23.

fig 16.13   *a photo-Darlington transistor*

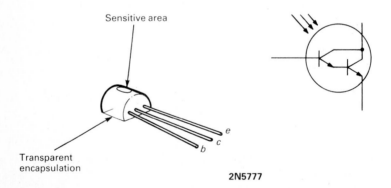

Sensitive area

Transparent
encapsulation

e
c
b

2N5777

Photosensitive ICs

A number of devices go further than the photo-Darlington, and combine a photosensitive junction with an amplifier and switching or shaping system. This, as usual with ICs, reduces the component count substan-tially with only moderate increase in the cost of the basic device. A typical example of the kind of thing that can be done is the LAS5V light-activated switch IC. This incorporates a high-gain amplifier and switching circuits that ensure a very rapid transition from the 'on' to the 'off' state and back again. Two external components are required, to set the thresholds for switching. The LAS5V provides an output that is directly compatible with CMOS or TTL logic circuits, or will even drive a small relay directly. Examples of both applications are given in Figure 16.14.

fig **16.14** *a photosensitive IC; this can be used to drive small relay or logic circuits*

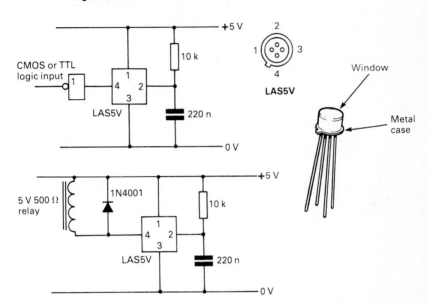

It should by now be clear that the general principle of photosensitivity can be applied to practically any semiconductor device all that is necessary is to expose the relevant junction to light. A range of photosensitive devices of a specialist nature—photothyristors and phototriacs, for example (see Chapter 17)—are available.

The most widely used devices are photoresistors, photo-Darlington transistors, and various light-sensitive ICs.

16.4 OPTO-ISOLATORS

The *opto-isolator* (sometimes known as an *opto-coupler*) combines the LED and photodiode or phototransistor.

Since both the LED and the phototransistor (or photodiode) can respond very rapidly to changes in applied current/light, it is possible to used a LED placed close to a phototransistor to provided an information link that is completely isolated electrically. The scheme is illustrated theoretically in Figure 16.15.

This kind of component is useful where circuits are at widely different potentials but need a signal to be passed from one to the other. The opto-isolator is a substitute for the isolation transformer in this application, but it is more efficient. A small opto-isolator in a DIL pack can easily

fig 16.15 *an opto-isolator*

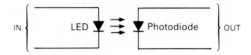

provide isolation between circuits at potentials differing by up to 4 kV. Opto-isolators are also valuable for feeding signals into sensitive circuits. If, for example, an electrically 'noisy' line (a line on which may be super-imposed high-voltage interference spikes) is required to feed an NMOS or CMOS circuit, directly connecting the two could well cause the sensitive NMOS or CMOS circuits to be destroyed by the interference. An opto-isolator provides the required coupling of the signal in complete safety.

The efficiency of an opto-isolator is stated in terms of the *transfer ratio*, which is simply the ratio of the output current to the input current, expressed as a percentage. A transfer ratio of 100 per cent would therefore provide an output current of 1 mA for each 1 mA of input current to the LED.

Opto-isolators using photodiodes would have transfer ratios of less than 5 per cent. Those using phototransistors are generally made with transfer ratios of 20 per cent, and those using photo-Darlington transistors can have transfer ratios better than 500 per cent—amplifying the input current by 5 times.

The frequency response is quite good, up to 200–300 kHz, according to type. Maximum currents are of course limited by the maximum rating of the LED and the maximum dissipation of the transistor.

16.5 PHOTOVOLTAIC CELLS

Commonly known as 'solar cells', photovoltaic cells convert light energy directly into electrical energy.

Early photovoltaic cells were very inefficient in terms of energy con-version, and produced very small currents and voltages. Since the 1960s intensive development has taken place, partly because it is thought that in some applications solar power, generated by means of photovoltaic cells, is a useful source of energy. This may yet prove to be true, but the most immediate use (and the reason for the large development budgets) is in the powering of space vehicles. In the inner solar system the sun represents a useful source of free energy for a space vehicle, an energy source that involves no payload of fuel, and will not run out in the anticipated lifetime of the space vehicle. Telecommunications satellites are invariably powered by photovoltaic cells, and the present designs are surprisingly efficient, converting about 20 per cent of the light energy falling on them into electrical power.

The structure of a silicon photovoltaic cell is shown in Figure 16.16.

An energy-level diagram for the *pn* junction exhibits the characteristic depletion region (Figure 16.17: refer to Chapters 6 and 7 if you need to) in which there are no holes or free electrons.

If, however, light is able to reach the junction, the energy absorbed from the light will break some of the bonds between adjacent atoms in the junction region, resulting in the formation of electron–hole pairs. These have to leave the depletion region immediately, and, if the two sides of the

fig 16.16 *a silicon solar cell, along with its circuit diagram*

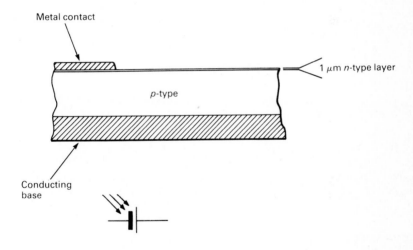

fig 16.17 *energy-level diagram for a pn junction*

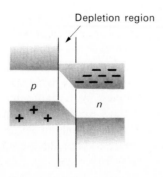

pn junction are connected through an external circuit—the load—they will flow through that circuit as an electric current (Figure 16.18).

Silicon photovoltaic cells are made with a very thin layer of *n*-type semiconductor on the 'top' layer, the *n* part of the *pn* junction. The layer may be only 1 μm thick, and is thin enough to allow incident light to penetrate to the junction region easily, resulting in a cell that is extremely efficient compared with older devices.

fig 16.18 *generation of electric current in the pn junction of a photovoltaic cell*

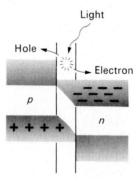

A silicon photovoltaic cell in the 'average efficiency' range, about 10 cm diameter, might well produce a current of 2 A in bright sunlight, with a potential of around 0.5 V. Since roughly 1 kW falls on every square metre of the earth's surface, this means that just under 8 W falls on the cell. With an output of 1 W, this means that the efficiency is about 12 per cent. Silicon photovoltaic cells are not cheap, and the most efficient types would be used only for defence and aerospace applications.

16.6 FIBRE-OPTIC SYSTEMS

Although not strictly electronic in nature, fibre-optic systems are important in communications applications. The fibre itself is a long 'wire' made of special glass or plastic. The outer layer has a different refractive index from the inner core, and if a light is shone on one end of the fibre, it will be transmitted down the whole length of the fibre to emerge at the other end with remarkably little loss of brightness. With careful fibre design and the correct wavelength of light applied to the end, attenuation can be reduced to as little as 30 dB per kilometre, for relatively inexpensive fibre. A suitable LED is used as the light source, and a photodiode as the detector. The system is shown in outline in Figure 16.19.

fig **16.19** *principle of a fibre-optic data-transmission system*

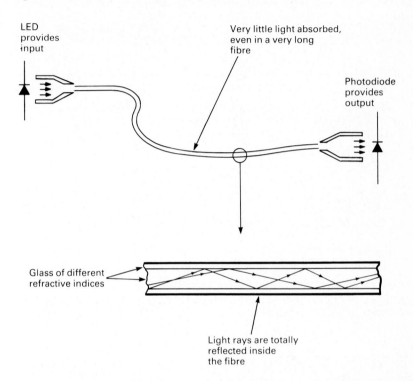

Very large bandwidths—up to 50 MHz even in inexpensive systems—can be obtained, more than can normally be obtained with conventional cables. Compared with copper cables, a telecommunications system using fibre-optics has many advantages. These include complete freedom from electromagnetic interference, complete electrical isolation, freedom from crosstalk (parallel cables inducing signals in one another, resulting in 'crossed lines'), low weight, greatly increased security (it is very difficult to 'tap' a fibre-optic phone line), and eventually cheapness (glass is basically a much cheaper material than copper).

The future of long-distance communication, apart from radio, seems to lie much more in fibre-optic systems than in transmission by means of electrical signals.

16.7 LASER DIODES

Although LEDs emit light over a narrow band of frequencies, the design of fibre-optic systems for use with LEDs is something of a compromise as the range of frequencies emitted by a LED makes it impossible to design the 'ideal' system. The reason for this is obvious if you consider a rainbow, and Newton's experiment with the prism – the amount by which light is refracted as it passes through a transparent medium depends on its frequency. This causes scattering, known as *chromatic aberration*, in any optical system involving refraction. A truly monochromatic light (a light consisting of a single frequency) can be focused to the theoretical limits of accuracy by a relatively simple lens, and the refractive indices of a fibre-optic system can be chosen to give near-perfect light transmission.

It is possible to generate monochromatic light at a high brightness level by means of a semiconductor LASER. The name LASER is another acronym, from **L**ight **A**mplification by **S**timulated **E**mission of **R**adiation. Lasers come in a whole range of sizes. The most powerful types have been used experimentally to shoot down aircraft in flight (although they are still a long way from the kind of thing you saw in Star Wars); semiconductor Lasers, even the biggest, are much more modest affairs, having a light output at most a few tens of times more intense than a LED.

Both the theory and the structure of a semiconductor laser are complex. Figure 16.20 shows, in diagrammatic form, a typical gallium arsenide laser diode. There are five layers of semiconductor. The central layer of *p*-type aluminium gallium arsenide has a thin strip of *p*-type gallium arsenide down the centre, and it is in this thin strip, about 1 μm by 10 μm, that the laser action takes place.

fig 16.20   *physical structure of a gallium arsenide semiconductor LASER*

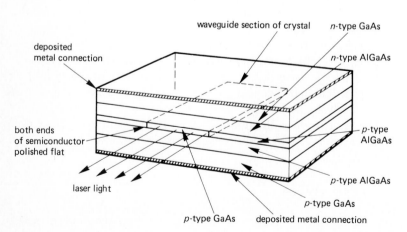

(NOTE   GaAs = gallium arsenide
        AlGaAs = aluminium gallium arsenide)

Without going into too much detail, it works like this. Light is generated at the *pn* junction in the normal way (see Chapter 7.8) and this light enters the thin strip of *p*-type gallium arsenide. Because the central strip has a higher refractive index than the material on either side, it behaves just like the fibre-optic light guide – light is trapped inside the strip. If the light in the strip is sufficiently intense, laser action begins. Photons are reflected backwards and forwards across the crystal, the ends of which are polished exactly flat and parallel. Some photons will collide with electrons in the conduction band. When this happens 'stimulated emission' can occur, releasing two photons travelling in exactly the same path. Both photons can now collide with electrons in the conduction band, and the result can be four photons, all in phase with each other and thus at the same frequency. The result is a rapidly increasing avalanche of photons, all trapped within the crystal. These photons eventually emerge from the end of the crystal as an intense beam of light at a single frequency. The light is also coherent, that is, the waves are all in phase with each other.

The monochromatic light source provided by the semiconductor laser is perfect for opto-electronic communications systems, and is also used in compact disc players (see Chapter 18.1) where it can be focused into a tiny circle less than 2 μm wide to 'read' the CD track.

QUESTIONS

1. Why is multiplexing often used in circuits operating LCDs?

2. When would a LED display be used instead of a LCD?

3. Suggest three circuits in which you might use a photoresistor.

4. When are opto-isolators used?

5. If solar cells can be used to power satellites and spacecraft for years at a time, why can they not be used to power cars?

CHAPTER 17

SEMICONDUCTOR AND ELECTROMAGNETIC DEVICES

This short chapter deals with a range of devices used in electronics, describing them briefly and outlining their functions. For various reasons the devices dealt with here are not covered elsewhere in the book, perhaps because they are outside the mainstream of electronics technology, or perhaps because they are components not generally regarded as 'electronic'. Nevertheless, this ragbag of items will prove of use and interest to the electronics engineer.

17.1 SEMICONDUCTOR DEVICES

Thyristors

The *thyristor*, or *silicon controlled rectifier* (SCR), is a component that has wide applications in the field of power control. In essence it is a very simple component, easy to use and easy to understand. The circuit symbol for the SCR is given in Figure 17.1. The circuit symbol is clearly reminiscent of the diode, and the SCR is indeed a form of diode. However, it will, under normal circumstances, fail to conduct current in either direction. The SCR can be made to withstand high voltages, and there are various designs that can be operated with a peak voltage of 50 V to tens of kV.

fig 17.1 *circuit symbol for a thyristor (SCR)*

If a voltage is applied to the SCR in such a way that, if it were a diode, it would conduct (forward-biasing it), and a small current is made to flow between the *gate* and *cathode*, the SCR will abruptly change from a non-conducting to a conducting mode, with characteristics similar to those of a forward-biased silicon diode. The forward voltage drop is generally higher, typically 0.7 to 1.3 V.

Turn-on takes place rapidly, within a few microseconds of the application of the gate current. Once turned on, the SCR will remain in the conducting mode, even if the gate current is removed.

Once triggered into conduction, the SCR will turn off again only when the current flowing through it is reduced below a certain value. This minimum current required to maintain conduction is called the *holding current*, and is between a few microamps and a few tens of milliamps, according to the type of device.

Thyristors are useful for controlling alternating current mains supplies, and are used in most lighting-control applications. For simple on/off circuits, the thyristor is placed in series with the circuit to be controlled. When the circuit is to be switched on, a gate current is applied; this principle is used in the photoelectric light control illustrated in Figure 16.11 (p. 244). An obvious limitation of this circuit is the fact that the a.c. supply is rectified—the lamp will be on at a lower brightness than usual, and some kinds of lamp or motor will not work at all. In the commercial design of Figure 16.11 a benefit is made out of this near-necessity—'extends the lamp life'—but in other circuits full control of alternating current is essential.

This can be obtained by using a bridge rectifier arrangement, so that the current is applied to the SCR in the proper sense for both positive and negative cycles of the a.c. waveform (see Figure 17.2). Circuits like this are possible because the thyristor is turned off every cycle, when the

fig 17.2 *the thyristor can be used to give full-wave control of an a.c. supply by the use of a bridge rectifier circuit*

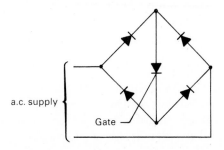

mains voltage crosses the zero line. With the gate signal removed, the SCR will turn off at the end of that half-cycle of the a.c.

The SCR is a power-control device, and as such it is made in a variety of sizes, from small devices capable of handling a few hundred milliamps at 50 V to huge units designed to deal with hundreds of amps at hundreds of volts. Small SCRs used in electronic circuits likely to be encountered by the engineer range from T092 encapsulated 300 mA versions up to about 40 A. Voltages are usually in the range 50-1200 V, with 400 V models preferred for control of mains electricity. It must, however, be emphasised that much larger devices are used in power applications. Gate currents— the minimum amount needed to trigger the SCR—range from a few hundred microamps for sensitive low-voltage SCRs, to a few tens of milli-amps for the larger ones.

Triacs

The *triac*, to give the more usual name for the *bidirectional thyristor*, is just what it sounds like. It is an SCR that will conduct in either direction, with the gate current flowing between anode 1 and the gate in either direction—a most flexible component. The circuit symbol is given in Figure 17.3.

fig 17.3 *circuit symbol for a triac*

The triac can be used as a direct substitute for the SCR in the circuit in Figure 16.11 (p. 244), with anode 1 in the same position as the SCR cathode. The gate resistor, controlling the gate current, may need to be reduced a little from 1.8 kΩ, as triacs tend to need rather higher gate voltages than the equivalent SCR. With the triac replacing the SCR, the circuit will switch the lamp on at full brightness.

Like the SCR, the triac will remain on, once triggered, until the current passing through it falls below the level of the holding current. When the triac is operated with mains electricity, this occurs once every half-cycle, or 100 times a second with a 50 Hz supply frequency.

Diacs

The *diac* is a specialised component, specifically designed for use with SCRs and triacs, though it has inevitably found its way into various other circuits. The 'long' name for the diac is the *bidirectional breakdown diode*, and its circuit symbol is given in Figure 17.4.

fig 17.4   *circuit symbol for a diac*

The diac normally blocks the flow of current in either direction, but if the voltage across it is increased to the *breakover voltage*, usually about 30 V, the diac begins to conduct. It does in fact exhibit a phenomenon known as *negative resistance*, for as breakover occurs the voltage across the diac drops by a few volts. It the diac is connected in a circuit in which a steadily increasing voltage appears across it, it will, at breakover, allow a sudden current 'step' to flow in the circuit. Figure 17.5 shows this in graphical form.

The diac is an ideal device for providing a suitable trigger pulse for an SCR or triac, and an applications circuit for a mains lamp brightness controller is given in Figure 17.6. This simple circuit is not recommended

fig 17.5   *a voltage and current for a diac*

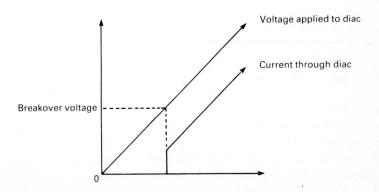

fig 17.6 *a functional circuit for phase control of an a.c. lighting circuit;*
this circuit will operate, but is likely to cause radio frequency
interference

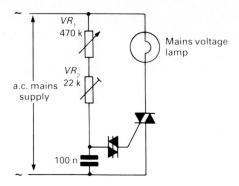

as a construction circuit, partly because mains voltages are present and
stringent safety precautions have to be taken (for example, most potentiometers have the shaft live to the brush, and thus live to mains), and partly
because it is electrically 'noisy' and produces radio interference. In commercial units—cheap because of mass production—radio frequency interference suppressors are fitted.

The circuit works by means of what is loosely known as 'phase control'.
As the voltage across the circuit rises, following the a.c. waveform, the
capacitor gradually charges up, at a rate controlled by VR_1 and VR_2. At a
certain point in the a.c. half-cycle the voltage across the capacitor will
reach the breakover voltage of the diac and the diac will apply a pulse of
current to the triac gate, triggering the triac into conduction and allowing
current to flow through the load. At the end of the half-cycle the triac will
switch off.

During the next half-cycle, the same sequence will occur, but with all
the polarities reversed (unimportant, because all the circuit components
will work with a voltage applied in either direction). Once again, the triac
will trigger after a certain delay, allowing a current flow through the load.

If the resistor VR_1 is set to a high resistance, the capacitor will charge
up slowly and the triac will be triggered near the end of each half-cycle.
If the resistor VR_1 is set to a low value of resistance, the triac will be triggered near the beginning of each half-cycle, applying almost full power to
the load. The amount of power flowing through the load is controlled by
VR_1, so the brightness of a lamp can be regulated over a wide range.
Because the circuit works by switching the power on and off, little heat is
dissipated and the control is efficient in terms of power used. VR_2 is
included in this circuit to set the maximum brightness level according to

the characteristics of the diac and triac; a production model would use a fixed resistor. This mode of power control is illustrated graphically in Figure 17.7.

fig 17.7 *illustrating the way in which the triac can be used with an a.c. supply to control the amount of power flowing through a load*

a.c. volts

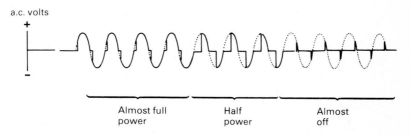

Almost full	Half	Almost
power	power	off

As we saw in Chapter 16 on optoelectronics, it is possible to make most semiconductor devices photosensitive just by using transparent encapsulations and optimising the design to allow light to fall on a sensitive junction. *Photothyristors* and *phototriacs* are therefore available, and these can be triggered in the normal way, or by ambient light rising above a certain level.

Unijunction transistors

Unijunction transistors (UJTs) are used primarily in oscillators. A typical unijunction oscillator is shown in Figure 17.8.

The UJT has three terminals, two *bases* and an *emitter*. The UJT in the circuit is an n-channel device; p-channel UJTs are available but uncommon (the circuit symbol has the arrowhead reversed for the p-channel

fig 17.8 *a unijunction oscillator*

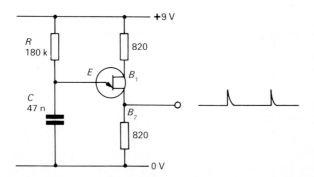

version). The UJT normally exhibits a high resistance between the two bases, about $10\,\text{k}\Omega$. If the emitter is at a low potential relative to B_1, the emitter-B_1 junction behaves like a reverse-biased diode, and negligible current flows through the emitter circuit. If the emitter potential is increased, to a point approximately equal to half the voltage between B_1 and B_2, the emitter-B_1 junction suddenly becomes forward-biased and the emitter-B_1 resistance falls to a very low value. In the oscillator circuit illustrated, the timing is controlled by R and C (similar to the circuit in Figure 17.6 above). When the UJT conducts, the capacitor is discharged very rapidly through the emitter-B_1 junction and the $820\,\Omega$ resistor. An output taken from B_1 therefore consists of a series of short pulses at the required frequency.

Thermistors

Used in temperature-sensing applications, *thermistors* are semiconductor resistors which change resistance with temperature. There are two types: negative-resistance temperature coefficient (NTC), in which the resistance decreases as temperature increases; and positive-resistance temperature coefficient (PTC), in which resistance increases with increasing temperature. Circuit symbols are given in Figure 17.9.

fig 17.9 *thermistors*

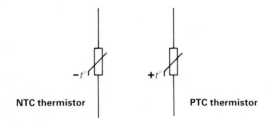

NTC thermistor   PTC thermistor

Voltage-dependent resistors

Voltage-dependent resistors (VDRs) are used in a few applications only. A VDR is simply a resistor whose resistance decreases as the voltage across the resistor increases. VDRs can be used to give a degree of voltage stabilisation in circuits operating in the same way as a Zener diode regulator. Because the VDR shows a gradual increase in conduction with increasing voltage, the regulation possible is not as accurate as can be achieved with a Zener. However, VDRs can be made with high resistance and high operating voltage, so they are still used in high-voltage regulating systems. The circuit symbol for a VDR is given in Figure 17.10.

fig **17.10** *circuit symbol for a voltage-dependent resistor (VDR)*

A VDR connected across a power supply will also provide a measure of protection from transient high-voltage peaks. The VDR is chosen to have a very high resistance at the supply voltage, but to have a low resistance at the high voltage levels associated with transients that might cause damage to the circuits the VDR is protecting.

Hall-effect devices

Named after the discoverer of the fact that magnetism could affect charge carriers (holes and electrons) in a solid, Hall-effect devices are semiconductors that react to an external magnetic field. These devices are sensitive not only to the existence of a magnetic field but also to its polarity.

Hall-effect devices are used for the measurement of magnetic fields, and also as switches, where they make a useful alternative to mechanical and optical switching systems.

A Hall-effect element is commonly included in a *Hall-effect switch*, in which an IC is used to provide positive on/off switching. An example is the TL172C, which is packaged in a standard T092 encapsulation. Normally the output is off, with a maximum current flow from the output of $20 \mu A$. When the magnetic field through the device reaches about $50 \, mT$ (milli-Teslas) it abruptly switches on, and up to $20 \, mA$ can be taken from the output.

17.2 ELECTROMAGNETIC DEVICES

Relays

Relays were among the first electrical components used as amplifiers, and were widely applied to early telegraph networks. A relay consists of an electromagnet—a number of turns of insulated wire wound on a soft iron core—that operates some sort of mechanical switch. Soft iron is used for the core because it does not retain magnetism easily. A simple relay is illustrated in Figure 17.11.

In a telegraph system, in which pulses of current are sent at a low rate of repetition, a relay can be used to amplify signals. The principle is shown in Figure 17.12.

fig 17.11 *a small changeover relay*

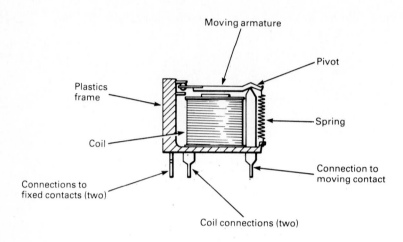

Moving armature

Pivot

Plastics frame

Spring

Coil

Connection to moving contact

Connections to fixed contacts (two)

Coil connections (two)

fig 17.12 *a relay used as an amplifier in a telegraph system*

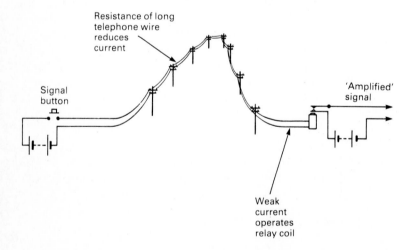

Resistance of long telephone wire reduces current

'Amplified' signal

Signal button

Weak current operates relay coil

There is no reason why the relay should not work more than one contact, nor is there any reason why contacts should not be normally closed
(to open when the electromagnet is energised) rather than the more usual
'normally open' form. Various circuit symbols are used for relays, and
Figure 17.13 shows two of them. Both relays are shown with two pairs of
contacts, one normally open and one normally closed. It is conventional to
draw the contacts in the positions they adopt with the coil unenergised.

fig 17.13 *circuit diagrams used for relays*

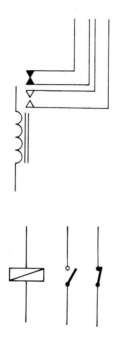

It is not always convenient to represent the relay contacts as neatly as in Figure 17.13 since a relay may have several sets of contacts controlling widely separated circuits; in this case the contacts are drawn in a convenient position, and are clearly labelled as belonging to a particular relay coil.

Relays come in an amazing range of sizes. At one extreme are huge machines in power stations and distribution systems, switching thousands of amps at thousands of volts. At the other end of the scale there are tiny low current relays that are encapsulated in a T05 can, for printed circuit use.

Reed relays

The magnetic reed relay is a fairly new device, developed for telecommunications before solid-state switching became possible. The reed relay was produced as a cheap and extremely reliable method of switching a low-power signal.

The reed consists of two springy blades of ferrous metal, sealed in a glass envelope full of inert gas. The blades are mounted so that they are not quite touching, about 0.5 mm apart. A reed switch is shown in Figure 17.14.

fig 17.14   *a 'single make' reed switch*

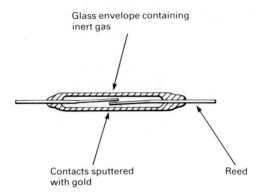

Glass envelope containing
inert gas

Contacts sputtered
with gold

Reed

If a magnetic field is brought close to the reed switch, the reeds become temporarily magnetised, and stick together, completing the circuit. When the magnetic field is removed, they spring apart again and break the connection. A reed relay is made simply by winding a coil round the outside of the reed switch. When a current flows through the coil, the induced magnetism makes the reed switch close.

The ends of the reed are coated with gold, and this, combined with the inert gas, prevents the contacts corroding. The reed relay is very reliable, and is fast in operation, closing in about 1–2 ms, and opening in less than 0.5 ms. Reed relays should not in general be used with inductive circuits, as arcing will damage the thin gold layer and may even weld the contacts shut.

Changeover reed switches are also common, and are constructed as shown in Figure 17.15.

fig 17.15   *a changeover reed switch*

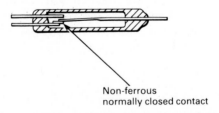

Non-ferrous
normally closed contact

Reed relays are commonly encapsulated in DIL packs, which makes them ideal for printed circuit use. There are still occasions when a relay is the best answer to a design problem, providing as it does complete isolation between circuits, and a capability of switching alternating or direct currents. Two DIL reed-relay pin diagrams are given in Figure 17.16.

fig **17.16** *pin-connection diagrams for two types of reed relay*

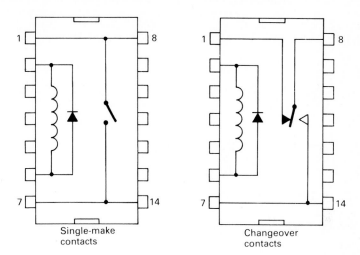

Single-make
contacts

Changeover
contacts

Solenoids

Where it is required to change electrical power into linear mechanical motion, a solenoid may be used. It is simply a coil wound round a suitable former, with a soft iron core free to move inside. When the coil is energised, the core is pulled into the coil by the magnetic field produced by the current flow. Solenoids are used to operate the mechanical parts of some cassette tape transport mechanisms, for instance. Solenoids are not particularly efficient in their use of power, and a large current is required if useful mechanical work is to be done.

Speakers

The speaker is, in reality, a solenoid. A typical speaker is illustrated in Figure 17.17.

A powerful magnet surrounds the *speech coil*, which is connected to the amplifier, etc., driving the speaker. A current flowing through the coil in one direction pulls the coil back into the magnet, whereas a current flowing through the coil the other way will push the coil out of the magnet. When the speaker is connected to the output of an audio amplifier, the movements of the coil will duplicate the waveform of the amplifier's output.

In order to produce a loud sound output, the coil has to move a reasonable volume of air, so the speaker *cone* is quite large, as large, in fact, as is practical. The movement of the cone, which has to be rigidly fixed to the coil, causes vibrations in the air—alternate compressions and rarefactions—which reproduce the sounds fed into the audio system.

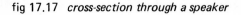

fig 17.17 *cross-section through a speaker*

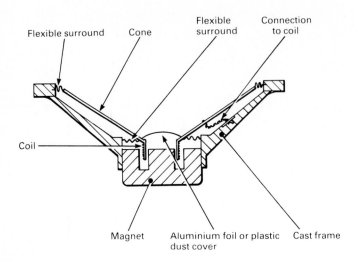

The design of speakers has been given a great amount of attention in the search for a speaker that will give the highest possible fidelity. Although a small speaker can reproduce the audio spectrum (the range of sounds that are audible to humans) reasonably well, it cannot provide the best possible fidelity across the whole range. Hi-fi systems therefore use two or three speakers, along with a simple electrical filter system to split the signal between the two or three speakers, according to frequency.

A bass speaker, for example (known as a 'woofer' in the hi-fi world), must move a lot of air to reproduce low bass sounds with sufficient volume. A large cone is necessary, or if the size requirement of the speaker cabinet does not allow this, then a small cone that has a large *excursion* (i.e. it moves in and out a long way) will suffice.

A treble speaker (known as a 'tweeter') does not have to move as much air, but in order to reproduce high frequency notes the cone must move very quickly. The large—and consequently heavier—cone of the bass speaker has too much inertia to move rapidly backwards and forwards, and so could not be used as a tweeter. Tweeters have small, very light-weight cones, with relatively little travel.

In both types of speaker it is necessary for the cone to be rigid; this is especially important in bass speakers. If the cone flexes as it moves, the additional vibrations that result will make the sound distorted.

Another factor is the amount of power that the speaker can (i) handle without distortion, and (ii) dissipate. All the output power of the amplifier is radiated by the speaker, most of it as heat (just a little as sound), and the coil must be capable of radiating the energy without melting. If a speaker is overloaded by a too-powerful amplifier, the cone may reach the limits of its travel, causing severe distortion.

The box, or *enclosure*, into which the speakers are fitted is also important for hi-fi work. The volume of air behind the speaker cone acts as a cushion for the cone, and damps its vibrations. Various reflections within the enclosure can also affect the sound, to a surprising extent.

Piezoelectric sounders

Where a 'bleep' is required, rather than an undistorted speech or music output, piezoelectric sounders may be used instead of small speakers. Applications are computer sound output systems, keyboard bleepers, small alarm bleepers and electronic alarm watches. The piezoelectric sounder is no more than a small piece of piezoelectric crystal, sometimes with and sometimes without an extra 'cone' (which, being flat, is referred to as a *diaphragm*). Low-voltage electrical signals applied to the crystal will make it flex, so a suitable audio frequency waveform will cause an audible output. Piezoelectric sounders are small (or very small), cheap, and not very loud.

Microphones

There are three main types of microphone in common use: *crystal*, *moving coil* and *capacitor*. The cheapest is the *crystal microphone*, which is very similar in design to the piezoelectric sounder. A simplified drawing of a crystal microphone is given in Figure 17.18.

Sound waves hitting the microphone will move the diaphragm in sympathy with them, flexing the diaphragm. The diaphragm moves the piezo-electric crystal, which generates an electrical output that is an analogue of

fig 17.18 *a crystal microphone*

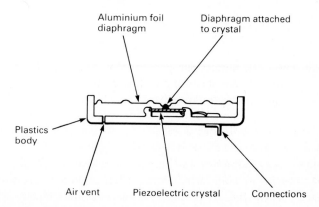

the sound waves. Crystal microphones characteristically have high voltage outputs—relatively speaking, for the crystal microphone may produce a few millivolts. The crystal microphone has a high internal impedance—several megohms—and so is not suitable for use with bipolar transistor amplifiers. FET amplifiers can be used, as these have high input resistances. The crystal microphone actually 'looks' to the amplifier like a capacitor of around 1-2 nF.

The output of the crystal microphone is not particularly linear, so it has been replaced in most common uses by the *moving-coil microphone.* Moving-coil microphones are made in a wide range of sizes and prices, from cheap models supplied with inexpensive cassette recorders to more expensive 'studio' microphones that have a very high audio quality. Figure 17.19 shows a simplified drawing of a moving-coil microphone. The construction is not unlike a speaker. Sound waves move the diaphragm, which is fixed to a lightweight coil. The coil is surrounded by a magnetic field, so as it moves a current is induced in the coil, which accurately mirrors the sound waves.

fig 17.19 *a moving-coil microphone*

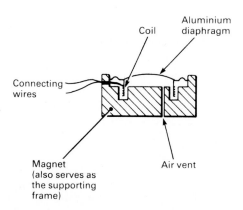

The coil impedance decides the impedance of the microphone. There are physical limits on how many turns of wire can be used, and for this reason most moving-coil microphones have a standard impedance of 200 Ω. This is suitable for matching both bipolar and FET amplifiers. The output of the moving-coil microphone is small, at most a millivolt. Moving-coil microphones are cheap and efficient, and can give good sound.

The most recent addition to the range of commonly used microphones is the *capacitor microphone*, which recent advances have brought out of the broadcast studio and into even the cheapest cassette tape-recorders.

The capacitor microphone utilises an *electret* element, which is a special capacitor with one 'plate' being a flexible diaphragm. If a voltage is applied to the electret element, the capacitor is charged, and there is no current flow apart from a negligible leakage current. However, sound waves hitting the diaphragm change the amount of capacitance (which depends, you may remember, on the spacing between plates). The amount of charge on the capacitor does not change, and as the capacitance varies so the voltage across the capacitor varies to maintain a constant charge. An integral IGFET preamplifier provides a current output and a typical impedance of 600 Ω.

Electret capacitor microphones give a high output compared with other types, and the sound quality is excellent. It is necessary to provide the electret element with a polarising voltage, to charge the capacitor; this is conveniently done with a 1.5 V battery, an HP7 size cell lasting more than a year. The smallest electret microphones are very tiny, powered with button-sized mercury or silver-zinc cells. A simplified diagram of a capacitor microphone is given in Figure 17.10.

fig 17.20 *a capacitor microphone*

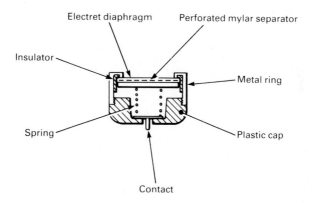

Meters

The last item under 'electromagnetic devices' is the meter. Although 'moving-needle' meters have been replaced in many applications by digital meters or bar graph displays, meters are still used in a variety of measuring and indicating applications.

Most meters are of the moving-coil variety, illustrated in Figure 17.21. The operation is straightforward and should be familiar. A voltage applied across the moving coil causes the moving coil to rotate on its bearings, against the returning pressure applied by the hairsprings. The coil will

fig 17.21 *a moving-coil meter*

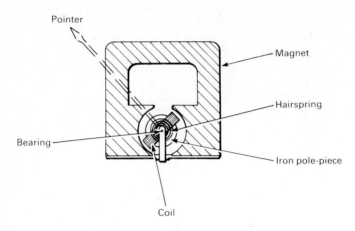

stop at a point where the increasing force of the hairsprings is exactly balanced by the magnetic forces. The more current passing through the coil, the further it will turn. A needle fixed to the moving coil moves along a scale, which can be calibrated according to the user's requirements. It is convenient to use the hairsprings to take the current to the coil. Other types of meter are seldom used in 'electronics' applications.

QUESTIONS

1. Draw the circuit symbol for a thyristor, and describe the important points about the device's operation.

2. Sometimes a VDR is used instead of a Zener diode for voltage regulation or limiting. Suggest two circumstances in which the VDR is to be preferred.

3. What advantages does a reed relay have over a conventional mechanical relay for telecommunications applications?

4. Why do hi-fi systems generally have more than one speaker in each enclosure?

5. Make a sketch of a circuit in which a relay is used to operate a triac. The triac is switching a mains lamp on and off. The relay provides isolation for the control circuit, which takes the form of a light-operated circuit, powered by a battery.

PART III
DIGITAL ELECTRONICS

INTRODUCTION TO DIGITAL ELECTRONICS

Digital electronics deals with the electronic manipulation of numbers, or with the manipulation of varying quantities by means of numbers. Because it is convenient to do so, today's digital systems deal only with the numbers 'zero' and 'one', because they can be represented easily by 'off' and 'on' within a circuit. This is not the limitation it might seem, for the *binary* system of counting can be used to represent any number that we can represent with the usual denary (0 to 9) system that we use in everyday life.

First, compare the way a digital system might be used to multiply two numbers together compared with a linear, or 'analogue', system. Let us imagine that we want to multiply 7 by 6 and arrive at the answer: 42. In an analogue system we could start with a voltage of exactly 7 V. If this were fed into an op-amp with a gain of exactly 6, the output voltage could be measured accurately and it would be found to be 42 V. This kind of 'computer' is quite feasible, but has limitations. Now suppose we want to multiply 7234 by 27 300. This is a different problem. Clearly it would now be impractical to use 1 V to represent 1, so we must use, say, 1 mV to represent 1. The input to the amplifier is thus 7.234 V. There now comes the problem of the amplifier to give a gain of 27 300—a massive design problem in itself. If it could be made, and made to work, the output would be 197 488.2 V, or nearly 200 kV! Even representing the input as 1 μV per 1, the output level would be almost 200 V—not really a practical solution, even supposing we could measure 200 V to the nearest microvolt.

A digital system, on the other hand, would treat the problem in a different way. Back to 7 times 6. Seven, in binary notation, is 111, and six is 110—see Figure 18.1 for a table giving comparable denary and binary numbers. We can represent a 0 by 0 V, and a 1 by, say, +5 V. Both six and seven could be represented using only three wires at one of two different potentials. The answer, 42, is represented in binary as 101010, which

needs six wires. To increase the capacity of the system to calculate 7234 ×
27 300 simply needs more wires; 7234 in binary is 1110001000010 and
27 300 in binary is 110101010100100. Hardly a readily recognisable or
convenient number for us, but it is still a way of representing a big number
without having to use electronic systems needing more than a few volts.
Also, the accuracy of the system in terms of reading voltage levels is not
required to be high; all that is needed is the ability to differentiate in the
circuits between 0 V and +5 V. The solution to the problem, 197 488 200,
is actually a rather long binary number, 1011110001010110111001001110,
28 binary digits, or 'bits', as the computer people call them. Although
binary numbers are long, the difficulty of manipulating them is nothing
compared with the problems associated with the equivalent analogue
systems. Rather than use 28 wires, a computer or calculator would use a
register holding the 28 'ons' and 'offs' in bistables, and the digits would be
processed in sequence rather than altogether, but such details are part of
computer science, not mastering electronics! Digital systems clearly have
big advantages.

fig **18.1** *some denary numbers with their binary equivalents*

Denary	Binary
0	0
1	1
2	10
3	11
4	100
5	101
6	110
7	111
8	1000
9	1001
10	1010
11	1011
42	101010
100	1100100

A circuit involved in storing and manipulating binary numbers is rather
complex, but in much the same way as a knitted sweater is complex, containing as it does a very large number of similar units. Digital systems
remained the province of the specialist until the advent of microelectronics.

Digital systems are perfect for making as integrated circuits. They have large numbers of regular circuit elements, they can be made without capacitors or inductors, and they can be made to use any convenient voltage. With mass-produced ICs containing complicated circuits available for the same price as a single transistor, it has now become possible to use digital techniques in all sorts of applications that were previously the province of analogue systems.

Take digital sound recording, for example. Hi-fi sound has long been associated with the very best of linear electronics. Tape-recorders in particular have undergone spectacular developments to bring them to the present quality and cheapness—but it is digital systems that are now used to make the 'masters' for records and tapes, in the best studios. Figure 18.2 shows a very small part of an audio waveform. This could be reproduced with very little distortion on the best tape-recorders available. But using digital recording techniques the waveform is *chopped* into brief time intervals (more than twice the highest audio frequency), and the voltage level measured at that point in time. The voltage level is assigned a numerical value—8 bits would give 256 different numbers—from zero

fig 18.2 *a digital recording technique*

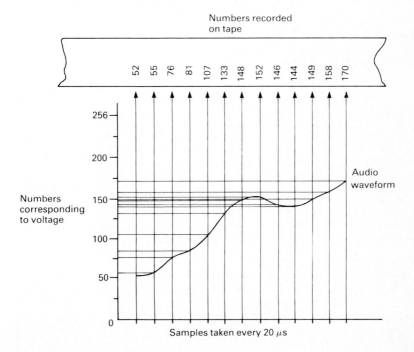

to maximum level. The numbers are recorded in binary form on a tape, at a rate (in this example) of somewhere over 300 000 bits per second.

To replay the tape, the numbers are turned back into voltage levels, and the original waveform reconstructed from the numbers. This may seem a complicated way of recording a concert, but the beauty of it is, if it works at all, it's perfect—or as near perfect as it needs to be.

A digital recording can be re-recorded over and over again, copied, re-copied, mixed, re-mixed, and if the numbers are readable (and they are read only as 'on' and 'off'), the music will be as perfect as the first recording, with no degradation whatsoever, and no background noise caused by successive re-recordings.

Digital recording is, it must be admitted, a rather 'high-technology' application of digital electronics. But digital systems are used in much simpler contexts, and there are surprisingly few different circuit elements in even the most elaborate system.

It is inevitable that digital electronic systems should lead to a study of computers and their organisation, for the digital computer is perhaps the greatest achievement of digital electronics technology. The following chapters, and the final section of this book, deal with digital circuit elements and complete systems, and, in the last two chapters, with microprocessors and digital computers. But you should remember that digital electronics covers not *only* computers but also a whole vast area of designs and ideas that make use of the ready availability of cheap digital integrated circuits.

18.1 DIGITAL AUDIO

One of the more impressive advances of digital electronics into the 'consumer' market has been the advent of the CD, or compact disc player. This is an entirely digital system that has provided a major advance in hi-fi audio reproduction. The CD system uses a 16-bit system, providing 65 536 different voltage levels, and a sampling rate of 44.1 kHz. Apart from the fact that nothing is lost between the original master recording and the millionth copy, the performance figures are, to the hi-fi enthusiast, quite compelling. Total harmonic distortion is 0.005%, compared with the best record at 0.15%. Signal-to-noise ratio is 90 dB, compared with the best record and player combination of 60 dB. Channel separation (that is, the amount of unwanted mixing that takes place between left and right stereo channels) is almost perfect at 90 dB, compared with a good record at 30 dB. Wow and flutter (variations in speed that affect the pitch or quality of the sound) are completely absent in the CD system.

On its own, recorded on tape, this kind of digital recording would be a big advance, but in the CD system it is combined with a new and better way of recording and playing back the digital information at the required minimum speed of almost a megabit (1 million bits) per second.

The compact disc is smaller than an LP record – only 120 mm diameter. It is played on one side only, but gives an hour's playing time. The recording is made in the usual spiral form, but starts from the middle, and is incredibly compact, with as many as 20 000 tracks across the 33 mm playing radius. The digital information is recorded as a series of tiny pits in a transparent plastic base, protected with a second layer of transparent plastic of different refractive index. A thin reflective layer of aluminium is also added. The diffraction effects give the CD its characteristic 'rainbow' appearance, and make it beautiful as well as efficient. Figure 18.3 shows the CD in section.

fig **18.3** *section through a compact disc*

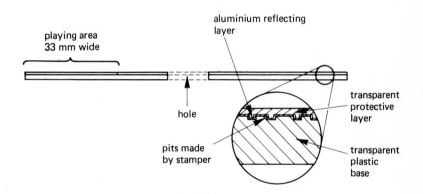

The pits are very small indeed, only about 0.5 μm in diameter and 0.2 μm deep, but are not easily damaged as they are protected by plastic on both sides. The pits are scanned by a focused light beam from a laser diode. A laser diode is used because the light from it is emitted at a single wavelength, is coherent (see Chapter 16.8) and is very directional, making it possible to focus it very sharply with a fairly simple optical system. The pits on the compact disc are always scanned at the same rate, so the disc has to spin faster when tracks near the middle are being scanned than when tracks near the edge are under the laser. The disc spins at a rate between about 200 and 500 revolutions per minute, over 4 million pits passing the laser every second. The minimum information rate mentioned above is easily exceeded, leaving extra capacity for all sorts of 'extras' like track number and time elapsed since the beginning of the disc.

Figure 18.4 shows a typical layout for the laser scanner. The laser 'pick-up' has to be guided very accurately, and all motions are feedback controlled. The distance between the scanner and the disc has to be exactly constant, the disc has to rotate at the right speed for any track, and – most difficult of all – the scanner has to follow the line of pits exactly even if the disc is not exactly concentric. Remember that the 'tracks' are only about 1.5 μm wide! Different manufacturers have developed different systems for controlling the scanner, but all involve high speed and extreme accuracy.

The complete lack of wow or flutter on the CD system may at first seem puzzling; after all, the disc is spun by a motor just like a turntable.

fig 18.4 *diagram of the optical system of the CD scanner*

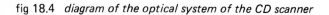

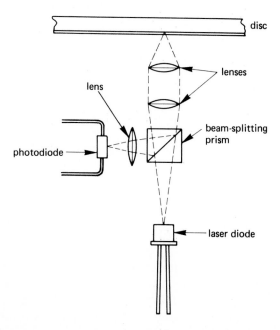

The reason is that minor variations in the disc speed will not affect the sound at all, for the disc is simply supplying data. The rate at which the data is translated into sound (the sampling rate) is controlled by the electronic circuits and is always exactly constant.

Unlike recording tapes, compact discs can be duplicated with a stamper, just like a gramophone record. The original is made using a high-powered laser to burn the pits into a special coating on a glass master disc. This is then used to make further masters from which the plastic discs are moulded.

The process is much quicker and more reproducible than recording on tapes, where there is always a balance to be struck between speed and quality of recording.

The electronic systems used in CD players are very complex and sophisticated, and could not have been developed for domestic use even five years earlier. In 1986 CD players represent, as they say, the 'state of the art' for this kind of product, well beyond the scope of this introductory book. But you have to begin somewhere, and the rest of this work deals with the basics of digital electronics and computing, from which everyday miracles like compact discs have developed.

LOGIC GATES

Take a simple, everyday thing like a coffee and soup vending-machine. Before it delivers a cup of coffee, two things must happen: (i) some money must have been put in; and (ii) the button for 'coffee' must have been pressed. We can write down the requirement as

money AND coffee button = coffee

and represent this diagrammatically—see Figure 19.1.

fig 19.1 *an AND logic system*

Both inputs (money and coffee button) must be present before there is an output. If we want to design a system that gives soup as an alternative, then we need a schematic like the one shown in Figure 19.2.

fig 19.2 *a logic system using two AND gates*

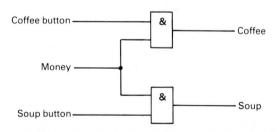

Notice that one of the inputs (money) is common to both gates. This, plus the other input, gives the relevant output. As far as it goes, the system is workable, but it has a design fault. Sooner or later someone will discover that if you press *both* buttons, you get coffee *and* soup. We could have predicted this by writing down a table of all possible combinations of logic inputs and their resulting outputs. Such a table is called a *truth table*, and a truth table for Figure 19.2 is given in Figure 19.3.

fig **19.3** *a truth table for a coffee/soup vending machine*

INPUTS			OUTPUTS	
Coffee	Money	Soup	Coffee	Soup
0	0	0	0	0
0	0	1	0	0
0	1	0	0	0
0	1	1	0	1
1	0	0	0	0
1	0	1	0	0
1	1	0	1	0
1	1	1	1	1

The last line is the one that shows up the problem. We can overcome our design problem by adding another stage to the logic, represented in diagram form by Figure 19.4.

fig **19.4** *a logic system for the coffee/soup vending machine*

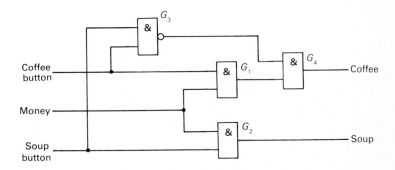

We should examine the workings of this in more detail. G_1 and G_2 are our original system; G_3 and G_4 have been added. The symbol used for G_3 is slightly different, the $\bigcirc$ on the output representing a negation. All this means is that the sense of the output is reversed—the output is always 'on' unless both inputs are also 'on' (in which case it is 'off'). Thus the upper input of G_4 is usually 'on', and when the lower input is also 'on'

(coffee button AND money) the 'coffee' output is also on. If both buttons are pressed, the output of G_3 goes off, suppressing the 'coffee' output. All you get is soup, and it serves you right for trying to cheat! A truth table for Figure 19.4 is given in Figure 19.5.

fig **19.5** *a truth table for the system shown in Figure 19.4*

INPUTS			OUTPUTS	
Coffee	Money	Soup	Coffee	Soup
0	0	0	0	0
0	0	1	0	0
0	1	0	0	0
0	1	1	0	1
1	0	0	0	0
1	0	1	0	0
1	1	0	1	0
1	1	1	0	1

You have probably recognised the diagrams as *logic diagrams*, and the systems shown can be implemented, exactly as they are, using digital integrated circuits.

19.1 AND/NAND GATES

Before going any further with the system, it is worth pausing to look at the logic gates that are commonly used in simple logic systems. Figure 19.6a shows AND and NAND gates. We met these in the coffee-machine logic, and their function should be obvious by now—but the truth table in Figure 19.6b formalises the functions. The output of the AND gate is 1 only when *both* inputs are also logic 1; the output of the NAND gate is 0 only when both inputs are logic 1.

19.2 OR/NOR GATES

Figure 19.7 shows the symbols and truth tables for OR and NOR gates. The output of the OR gate is 1 if *either* of the inputs is 1, or if *both* of them are 1. In the same way, the output of the NOR gate is 0 if either or both of the inputs are 1.

fig 19.6

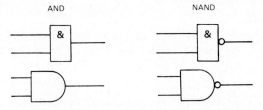

AND

NAND

(a) *logic symbols for AND and NAND gates; the upper symbols are the recommended IEC/British Standard symbols, while those below are the American standard; both types of symbol are used, the American version being the more common*

AND

Inputs		Output
0	0	0
0	1	0
1	0	0
1	1	1

NAND

Inputs		Output
0	0	1
0	1	1
1	0	1
1	1	0

(b) *truth tables for the AND and NAND gates*

fig 19.7 *OR and NOR gates, along with their truth tables; once again, the upper symbols are IEC/BS, the lower ones the US standard*

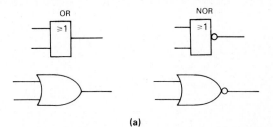

OR

NOR

(a)

OR

Inputs		Output
0	0	0
0	1	1
1	0	1
1	1	1

NOR

Inputs		Output
0	0	1
0	1	0
1	0	0
1	1	0

(b)

19.3 **NOT GATE, OR INVERTER**

Figure 19.8 shows the simplest logic element (no truth table necessary!), the NOT gate, more usually called an *inverter*. It simply changes the sense of the input, 0 output for 1 input and vice versa.

fig **19.8** *a NOT gate (or inverter); IEC/BS symbol on top, US symbol below*

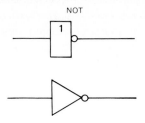

19.4 **EX–OR GATES**

Figure 19.9 shows an Exclusive–OR gate, along with its truth table. The Exclusive–OR gate is just like an ordinary OR gate, *except* that the output is only 1 when just one of the inputs is 1. If more than one input is at logic 1, or if all inputs are at logic 0, the output is 0. EX–OR gates are less commonly found than the AND/NAND, OR/NOR varieties.

fig **19.9** *an exclusive-OR gate, with its truth table; IEC/BS symbol on the left.*

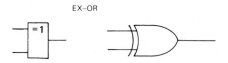

Inputs		Output
0	0	0
0	1	1
1	0	1
1	1	0

19.5 **MORE COMPLEX GATES**

It is possible to ring the changes on the basic gates. Gates of all types often have more than two inputs—Figure 19.10 shows a CD4068 eight-input NAND gate. If one or more inputs are at logic 1, the output is 0. If all the inputs are 0, the output is 1.

fig **19.10** *an 8-input NOR gate, IEC/BS symbol. From now on, throughout this book, IEC/BS symbols are used; the equivalent US symbol is easily derived by comparison with Figures 19.6 to 19.9*

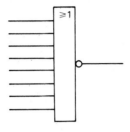

Gate *inputs* can also be negated, reversing their sense. This could be done using an inverter in the input line (Figure 19.11*a*); or if the inverter is part of the same IC as the gate, it can be drawn like the one in Figure 19.11*b*.

fig **19.11** *a NAND gate with one input inverted—two alternative representations*

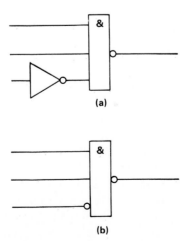

(a)

(b)

19.6 MINIMISATION OF GATES

It is only sensible to design any logic system with the smallest possible number of gates, all else (the price of different types of gate) being equal. For simple systems it is usually possible to minimise the number of gates by inspection; for more complicated systems boolean algebra can be used to formalise the problem and arrive at a solution (see N. M. Morris, *Digital Electronic Circuits and Systems*, Macmillan, 1974).

Clearly, it takes only a short time to realise that the system in Figure 19.12*a* can be reduced to the system in Figure 19.12*b* without any change in its functioning. It is less obvious that Figures 19.13*a* and 19.13*b* are functionally the same—but it is obvious if you draw up the truth table for them.

fig 19.12

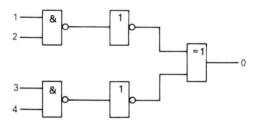

(a) *a logic system*

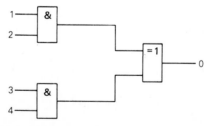

(b) *a simpler version, performing the same function*

Figure 19.14 shows a simple logic system that was actually used in a cassette tape-recorder intended for use with a microcomputer. F_1 and F_2 are inputs controlled by the computer, to stop and start the recorder motor. The two gates on the right form a *bistable* (see Chapter 21) which 'latches' the motor on or off. Switch '*M*' is for manual operation, and turns the motor on by applying a logic 1 pulse to the gate.

At first sight the circuit can be simplified by avoiding the double negation in the NOR gate and inverter, replacing them both with a single OR

fig 19.13

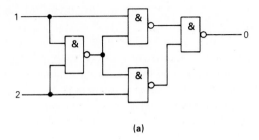

(a)

(a) *a logic system*

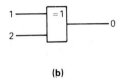

(b)

(b) *an exclusive-OR gate performing the same function*

fig 19.14 *a bistable control for a tape drive motor*

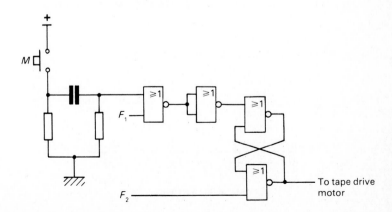

gate, as in Figure 19.15. So why would a sensible (and cost-conscious) manufacturer prefer the more complicated version in Figure 19.14? The answer lies in Figure 19.16, which shows the pin-out of the SN74LS02 integrated circuit.

This integrated circuit includes four NOR gates (it is known as 'quad 2-input NOR gate'), which is just right for the system in Figure 19.14. You cannot obtain a single integrated circuit that incorporates an OR gate plus NOR gates, so the apparently simpler version in Figure 19.15 actually requires two integrated circuits, which in turn means more circuit-board space, more soldered connections, and higher assembly costs. In almost all

fig **19.15** *a simpler but more expensive version of the circuit given in Figure 19.14*

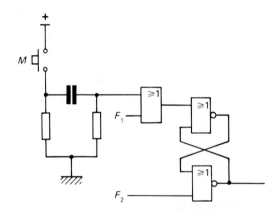

fig **19.16** *the pin connections of the SN74LS02 quad NOR gate*

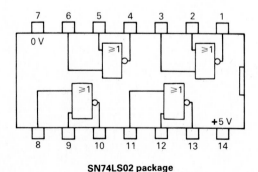

SN74LS02 package

practical systems, minimisation of *components* and *cost* is the requirement. Reducing the number of gates in a given system may not always further that end.

Where large systems are involved (e.g. counters, clocks and computing circuits) it is almost always economic to use a large-scale integrated circuit (LSI) or even a microprocessor, despite the fact that only a fraction of the device's total capacity may be used. Very complicated LSI circuits are relatively cheap—a Z80 microprocessor chip costs less than twenty times as much as the simplest gate.

Having covered the basic types of gate, in the next chapter we shall consider the two major kinds of technology that are used in the implementation of integrated circuit logic elements.

QUESTIONS: CHAPTERS 18 AND 19

1. Why is the binary system of numbers used in computers?

2. Sketch a logic system in which the following truth table is implemented:

INPUTS			OUTPUTS	
1	2	3	1	2
0	0	0	0	1
0	0	1	0	0
0	1	0	0	0
0	1	1	1	0
1	0	0	0	0
1	0	1	0	0
1	1	1	1	1

3. Draw circuit symbols for (i) a 3-input NAND gate, (ii) an inverter, (iii) a 4-input OR gate with one negated input.

4. What is the largest (denary) number that can be represented by eight binary digits?

5. Draw a logic circuit including one EX-OR gate and one NAND gate, any connections. Now construct a truth table for the circuit.

LOGIC FAMILIES

Although there are many ways of making integrated circuits, and several technologies are available to the manufacturer, only two major 'families' of general-purpose logic ICs have developed. The first, and oldest, family is TTL. TTL stands for 'transistor–transistor logic', this logic family having replaced the older (now obsolete) DTL—diode–transistor logic. The second logic family is CMOS (pronounced 'sea-moss') which stands for 'complementary metal oxide semiconductor'. Both families include a very large number of different devices: gates, flip-flops, counters, registers and many other 'building-block' elements. In many cases there are direct substitutes, but, as we shall see, it is not usually possible to mix devices from the two families.

Both groups have their own special advantages and disadvantages, and when a designer is working on a digital system the choice of logic family will probably be one of his first decisions.

20.1 TTL

This family is based on the bipolar junction transistor, and was perhaps the first commonly available series of logic elements. Although the logic designer rarely needs to know how the inside of an IC works, it is instructive, when comparing logic families, to look at the workings of a simple gate. We can take as an example a 2-input NAND gate, numbered, appropriately enough, SN7400 in the TTL series. Figure 20.1 shows the circuit of this gate.

Notice that the circuit has a familiar 'transistor amplifier' look about it, except that the circles round the transistors have been left out—they represent the encapsulations. The input transistor is also unusual, by discrete component standards, in that it has two emitters. The circuit operates as follows. TR_2 is normally off (both inputs logic 0), which means that TR_3 is on, the 1.6 kΩ resistor providing the base current. TR_4 is off, its base

fig 20.1 *a typical bipolar 2-input NAND gate*

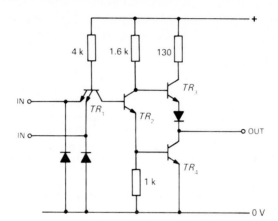

being clamped to 0 V by the 1 kΩ resistor. The output is therefore at a potential just below that of the positive supply, or logic 1. If *both* inputs are taken to logic 1, then a base current is provided for TR_2, and TR_3 switches off—its base is taken to 0 V via TR_2 and the base–emitter junction of TR_4. TR_4 switches on. The output is now at a potential near that of the supply 0 V, or logic 0.

It is clear that moderate amounts of power are required, since either TR_3 or TR_4 has to have a base current flowing the whole time. The SN7400 requires about 2 mA per gate at its operating voltage of 5 V—not a lot, but enough to make power supplies a headache in complex systems and enough to rule out battery operation for all but the simplest systems. It is equally clear that there is a definite limit on the amount of current each gate can provide; it varies from device to device, but the SN7400, which can provide at its output around 400 μA at logic 1 and 16 mA at logic 0, is typical. There is a big difference between the current that a TTL IC can *source* (i.e. when it is at logic 1) and *sink* (i.e. when it is at logic 0). Practically, this is of no importance when interconnecting gates since TTL inputs present a very high resistance when looking at a logic 1. It is of importance when *interfacing* TTL with other systems and devices, as we shall see later.

Fan-out

Manufacturers quote the capability of the output of a TTL circuit in terms of the number of standard TTL inputs it can successfully drive. This figure is termed the *fan-out*. For SN74-series TTL ICs, the fan-out is usually 10: each output can drive up to ten inputs.

Operating speed

TTL gates work very quickly. The SN7400, for example, takes just fifteen nanoseconds to change state. TTL counters—which we shall be meeting in the next chapter—can count at frequencies up to 50 MHz, and the faster LS series (see below) at up to 100 MHz.

Power supplies

TTL logic circuits require a standard power supply of 5 V. The amount of tolerance is not large, and the limits of reliable operation are between 4.5 V and 5.5 V. It is therefore imperative that some form of power-supply regulation be used. Power-supply circuits form an interesting and quite important branch of electronics in themselves, and the design of a good power supply used to be very difficult. Fortunately, we now have cheap and extremely reliable power-supply *regulator ICs* which greatly simplify the job. Figure 20.2 shows a supply suitable for use with a 9 V battery (the ubiquitous PP9), capable of providing up to 100 mA.

fig 20.2 *a power supply for operating small TTL logic systems*

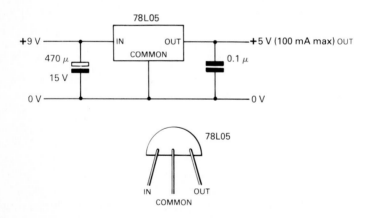

The regulator IC, the 78L05, has a bigger brother, the 7805, which can supply up to 1 A. Regulators to provide up to 5 A at 5 V are commonly available. Since this represents 25 W (5 × 5), and the power is all dissipated by the ICs, it is clear that there is a cooling problem with large TTL systems! Fan cooling is quite commonly used.

A complete mains power supply for TTL circuits, capable of delivering up to 1 A at 5 V, is illustrated in Figure 20.3. The 7805 is protected against overheating and being shortcircuited, and even when used with experimental systems is quite hardy. Needless to say, the mains part of this unit must be effectively insulated.

fig **20.3** *a power supply for operating more complex TTL logic systems*

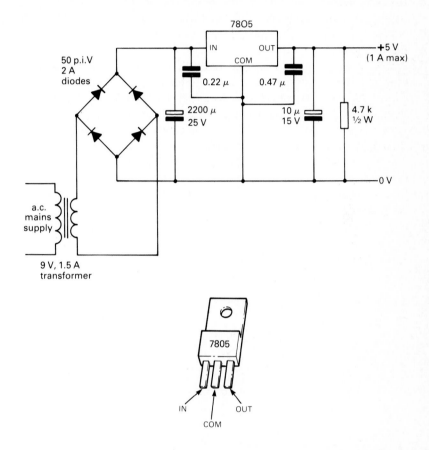

A TTL system can impose very heavy instantaneous loads on a power supply. This can in turn result in interference 'spikes' in the power lines, upsetting the normal working of parts of the system. A cure can usually be effected by connecting small capacitors directly across the power lines, as near as possible to the ICs themselves. Usually $0.1\,\mu\text{F}$ ceramic capacitors are recommended, but $10\,\mu\text{F}$, $15\,\text{V}$ tantalum 'bead' capacitors can some-times prove even better. Whichever type is used, about one capacitor per five ICs seems to work well.

Unused inputs
It is important to ensure that unused inputs of TTL gates are connected, either to *used* inputs (but remember the fan-out limit of the preceding device), to the positive supply, or to supply $0\,\text{V}$. Connections to $5\,\text{V}$

should be made through a 1 k resistor; you can connect several inputs to the same resistor. Failure to terminate unused inputs may lead to the gate oscillating, harmful because it may cause the circuit to overheat.

Encapsulations

Although many encapsulations are listed by the makers, the DIL package (dual-in-line) is practically universal. A 14-pin DIL package is illustrated in Figure 20.4. The pins are always numbered from pin 1, which is marked with a notch, or is the pin to the left of a notch in the end of the package, with the notch to the top. The pin numbers run in an anticlockwise direction, looking at the top of the package (which is held 'legs down'). The DIL package is usually made of plastic, though some high-quality devices may have ceramic packages, which are much more expensive.

fig 20.4 *a 14-pin DIL package*

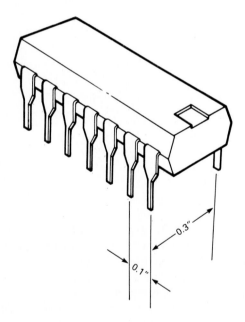

Schottky TTL

TTL circuits are being improved continually, and a major advance has been in the introduction of a range of *low-power Schottky TTL* circuits. They use the same code numbers as the standard series, but with 'LS' before the type code—e.g. SN74LS00. Operating speeds are about twice as high, and power consumption is as low as one-fifth of that of standard TTL. The only 'minus' is the cost, about 20–30 per cent more than the standard range.

Mainly because of the much reduced power consumption, the Schottky TTL is now more widely used than the standard TTL. In any digital system, and particularly in a large one, power supplies represent an appreciable fraction of the total cost, and the savings that can be made here will more than pay for the higher cost of the 'LS' series circuits. Indeed, it is the problem of power supplies that has been an important factor in the rapid growth and acceptance of the other major logic family, CMOS.

20.2 CMOS

Whereas TTL ICs are based on the bipolar junction transistor, CMOS ICs are based on the field-effect transistor. Figure 20.5 illustrates a CMOS 2-input NAND gate. Before looking at the way this circuit operates, bear one fact in mind: the MOSFET is a voltage-operated device, and the gate insulation means that, for all practical purposes, there is no gate current. It is this that enables CMOS ICs to achieve astonishing power economies.

fig 20.5 *a 2-input NAND gate using CMOS technology*

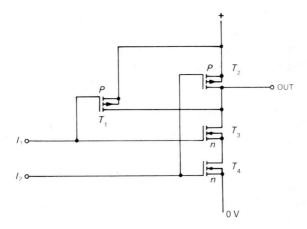

Clearly, therefore, the inputs for the NAND gate in Figure 20.5 will take no current. If the output is feeding another CMOS gate, then the output will not be required to deliver any current!

The operation of the gate shown in Figure 20.5 is as follows. Consider the circuit with both inputs taken low (logic 0). With I_1 low, T_3 will be off (high resistance) and T_1 will be on. With I_2 low, T_4 will be off, and

T_2 will be on. This much should be quite clear—if it is less than clear, you should re-read Chapter 9. The output terminal is held at a high potential (logic 1) via T_1 and T_2.

Now let us take I_1 to logic 1. This will switch on T_3 and switch off T_1, but the output will remain at logic 1 because T_2 is still on and T_4 is still off. The same thing happens if I_2 and not I_1 is taken to logic 1. If, however, both inputs are at logic 1, then T_1 and T_2 are both off, isolating the output from the positive supply, whereas T_3 and T_4 are both on, making the output logic 0. Obviously, the simple circuit of Figure 20.5 functions effectively as a NAND gate. It is a characteristic of CMOS that most power is used during the transition from one state to another, so the average power required is a function of the switching frequency. Typically, a NAND gate like the one in Figure 20.5 would use less tha $0.1\,\mu$A with a 5 V supply when working!

CMOS has some other advantages. It can be used with a wide range of power-supply voltages—the RCA 'B' range, for example, works from 3 to 18 V. Because of the lack of any appreciable voltage drop across the IGFETs when they are turned on, the logic 1 and logic 0 outputs are very close to the power-supply voltages—only about 10 mV below the + supply or above 0 V. CMOS logic also switches neatly at half the supply voltage. For a 5 V supply, the output switches as the input crosses 2.5 V; for a 12 V supply this happens at 6 V; and so on.

Finally, CMOS has far greater immunity to power-supply 'noise' than TTL and will tolerate noise of at least 20 per cent of the supply voltage without problems. Clearly, CMOS is the first choice for battery-operated equipment. A complex CMOS system can be run from a small 9 V battery, and the battery will have a very long life; the voltage can drop to below 6 V before the circuits are affected adversely. No power supply or smoothing circuits are generally required, though a small ($10\,\mu$f or so) capacitor is often connected across the battery. In general, power-supply decoupling capacitors such as tend to be sprinkled liberally around TTL circuits are not needed for CMOS.

But there are disadvantages, too. CMOS is slower than TTLs. A CMOS counter might typically work at up to 5 MH—quite quick, but only a tenth as fast as a corresponding TTL circuit. The available output current is also low. This is not, of course, a problem within a wholly CMOS system (fan-out of a CMOS gate is more than 50) but means that buffering circuits are generally needed to drive indicators or control relays. The output is equivalent to a series resistor of 400 Ω–1 k, so by Ohm's Law we can see that with a 10 V supply only a few milliamps can be taken from the output. In practice, the output current is further limited by the safe dissipation of the device, around 200 mW per *package* (i.e. 50 mW per gate for a quad-NAND gate like the CD4011).

Handling CMOS

As we saw in Chapter 9, the MOSFET is susceptible to damage from electrostatic voltages. The same is true of the CMOS IC, though the manufacturers build in various safety features. A CMOS input actually looks like Figure 20.6. This protection system works for voltages up to 800 V–4 kV depending on the device. But since your body can easily pick up a charge of 10 kV when you walk across a nylon carpet, precautions are still essential. A suitable bench top for working with CMOS is made of a conductive material and is earthed.

fig **20.6** *typical protection circuit for CMOS inputs*

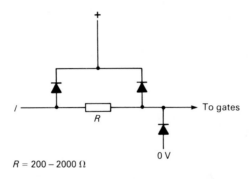

$R = 200 - 2000 \ \Omega$

High electrostatic potentials can also appear on the bit of a soldering iron. A rather inconvenient cure is to earth the bit (but not through the earth pin of the power socket). A better cure is to use a soldering iron with a ceramic shaft—the ceramic conducts heat from the element to the bit of the iron, but with very high electrical resistance and low capacitance.

The use of a ceramic-shafted soldering iron and earthed conductive bench top should provide sufficient electrostatic protection for all modern CMOS devices.

20.3 COMPARISON OF TTL AND CMOS

It is useful to compare the main characteristics of the two major logic families. The circuit designer will generally choose the logic system according to the design requirements—power supply (mains or battery?), operating speed, outputs required, and of course price.

Figure 20.7 compares a simple gate—a quad-NAND gate in TTL (LS) and CMOS; and Figure 20.8 compares a more complicated circuit—a 4-bit binary up/down counter.

fig 20.7 *comparing characteristics of similar simple TTL and CMOS circuits*

		Supply voltage	Supply current per gate	Fan-out	Input current (logic 0)	Short-circuit output current		Typical maximum frequency
						logic 0	logic 1	
TTL	SN74LS00	5 V	400 μA	20	400 μA	30 mA	5 mA	45 MHz
CMOS	CD4011	3 – 18 V	80 nA	>50	10 pA	3 mA		5 MHz

Quad – NAND gate: comparison of characteristics; TTL and CMOS

fig 20.8 *comparing characteristics of similar complex TTL and CMOS circuits*

		Supply voltage	Supply current	Minimum input pulse duration	Typical maximum count frequency
TTL	SN74LS193	5 V	17 mA	20 ns	25 MHz
CMOS	CD4029	3 – 18 V	2.5 μA	200 ns	3 MHz

4-bit binary up/down counter: comparison of characteristics; TTL and CMOS

QUESTIONS

1. The output of a TTL gate is connected, via a meter that reads current, to the 0 V supply line. Suggest typical current readings when the output is at (i) logic 1, (ii) logic 0.

2. What are typical power-supply requirements for (i) TTL, (ii) CMOS?

3. Compare the TTL and CMOS logic families for the following characteristics: (i) current consumption from the supply, (ii) speed of operation, (iii) immunity from interference, (iv) tolerance of variations in the supply voltage.

4. What precautions need to be taken when handling CMOS circuits?

5. Which logic family would you use in designing a system for the following uses: (i) a stop-watch, (ii) high-speed data transfer system for a computer, (iii) control logic for electronic ignition system in a boiler?

COUNTING CIRCUITS

Counting circuits are among the most important of all sub-systems, and are used in a tremendous variety of electronic equipment. People unused to electronics often fail to appreciate just how fast an electronic counting circuit can work. A typical TTL counter might accept a count frequency of 30 MHz. You only begin to realise the speed of this when you compare it with other things that are normally considered very fast. Imagine you have a revolver, and that a counter working at 30 MHz is connected in such a way that it begins to count the instant the firing pin hits the cartridge. Fire the pistol and watch the counter—if you have good eyesight you will notice that the counter reaches a count of around 10 000 *before* the bullet emerges from the end of the pistol's barrel! Electronic counters can also work slowly, but the potential for such rapid counting opens the way for all sorts of useful designs.

Before the advent of integrated circuits, counting circuits were rather complicated and difficult to build. Electronics technicians and designers needed a detailed knowledge of many types of counting circuit, mostly based on the transistor bistable configuration, but some based on special 'counting tubes'. For low-speed operation, mechanical counters worked by electromagnets were often the best answer.

The position is very different nowadays. Many types of counter are available in integrated circuit form, and it is not really essential to know in detail what goes on inside the circuit. Indeed, manufacturers with an original circuit design are often reluctant to publish details. What we *do* need is a clear idea of the various forms of counter, what their limitations are, and how they can be used.

21.1 THE BISTABLE

The bistable is one of the simpler circuit elements. Before the integrated circuit era, the bistable circuit had to be made up from discrete components, and required, in its simplest form, two transistors and four resistors. The circuit is given in Figure 21.1.

Imagine that TR_1 is conducting, so that TR_1 has a very low resistance. Point 'X' in the circuit will be near to 0 V (disregarding the forward voltage drop of the junction of the transistor). The base of TR_2 is held at this level via R_2, and since the transistors in this circuit are of the *npn* type, TR_2 is turned off. Point 'Y' is therefore at a high potential, and this high potential is applied to the TR_1 base, keeping TR_1 turned on. The circuit is completely stable in this condition and will remain indefinitely with TR_1 on and TR_2 off.

fig **21.1** *the simplest transistor bistable requires six discrete components*

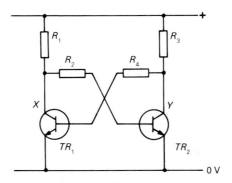

But the circuit is symmetrical. Both 'sides' are identical, and it will be equally stable with TR_1 off and TR_2 on (hence the name 'bistable').

The bistable can be made to switch states by the application of a suitable voltage to the base of one or other of the transistors. If TR_1 is on, a brief pulse of positive voltage to the base of TR_2 will turn on TR_2 instead, switching TR_1 off. A simple demonstration circuit is easy to make, and Figure 21.2 gives all the information needed. LEDs are connected in the two collector circuits to show which transistor is switched on. Momentary connection of terminal T_1 or T_2 to the positive supply will switch on the associated transistor, and switch off the other one.

More simply, bistables can be made using NOR gates. The circuit shown in Figure 21.3*a* (p. 300) performs in the same way as the discrete component version in Figure 21.2, the section shown dashed is required only in the demonstration circuit to indicate which state the circuit has adopted. Once again, it is quite simple to build the demonstration circuit. Details are given in Figure 21.3*b*.

fig **21.2** *a practical transistor bistable, suitable for demonstration purposes; component tolerances are uncritical, and most npn transistors will work satisfactorily in this circuit (red LEDs make useful indicators to show which transistor is on)*

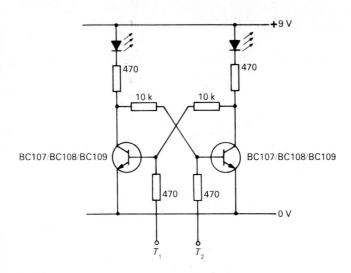

21.2 INTEGRATED CIRCUIT FLIP-FLOPS

Purpose-built bistable ICs are also available. A bistable similar to that in Figure 21.3a is termed *an S-R flip-flop* (the *S-R* standing for SET-RESET). The circuit symbol for this is given in Figure 21.4.

When S is 1 and $R = 0$, Q is also 1, and $\overline{Q}$ is 0. When S is 0 and R is 1, the circuit changes state and Q becomes 0 and $\overline{Q}$ becomes 1. If both S and R are 0 or 1, the circuit will remain in the state in which it was last set. Put simply, a logic 1 on either the S or R input makes the circuit flip (or flop) over to the Q or $\overline{Q}$ output respectively.

The third input is the *clock input*, *CK*. If both S and R are at logic 1, a 1 pulse applied to the *CK* input will cause the circuit to change state (Q and $\overline{Q}$ becoming 0 and 1 or 1 and 0). Each successive clock pulse—a swing from 0 to 1 and back again—will make the circuit change state. It is usual (but not universal) for the circuit to change state as the clock pulse falls from 1 to 0. Figure 21.5 shows the clock input, and the way the output changes state.

The *CLR* (clear) input has to be held at logic 1 for normal operation. It is taken to 0 to reset the bistable, which makes Q go to logic 0, $\overline{Q}$ to 1, and suppresses all further changes of state. The various combinations of inputs and outputs are shown in the table in Figure 21.6.

fig **21.3**

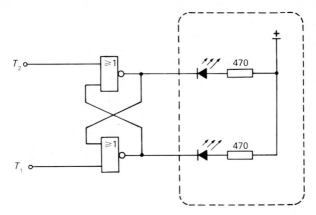

(a) *a bistable constructed with NOR gates*

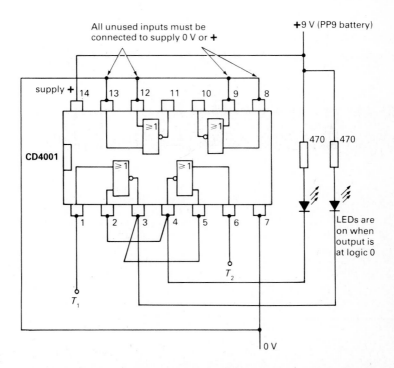

(b) *a demonstration NOR gate bistable, with LED indicators to show the state of the circuit: a CD4001 integrated circuit is used, and a 9 V PP3 battery provides the power source*

fig 21.4 *S–R flip-flop symbol*

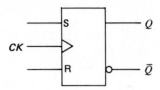

fig 21.5 *clock and output waveforms of the flip-flop*

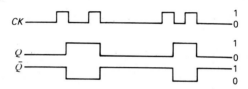

fig 21.6 *function table for the S–R flip-flop*

Inputs				Outputs	
CLR	CK	S	R	Q	$\bar{Q}$
0	X	X	X	0	1
1	⊓	0	0	Q_0	$\bar{Q}_0$
1	⊓	1	0	1	0
1	⊓	0	1	0	1
1	⊓	1	1	TOGGLE	

X = doesn't matter

$Q_0, \bar{Q}_0$ = outputs remain in the state they were in before the indicated input conditions were established

TOGGLE = Q and $\bar{Q}$ change state

⊓ = clock pulse rises from 0 to 1 and falls to 0 again

Figure 21.7 shows another more complicated type of bistable called a *J-K master-slave flip-flop*. This consists of a pair of *S-R* flip-flops, connected together by various logic gates. The input gates are connected to the outputs, so that the *J* and *K* outputs are in the opposite state. This ensures correct operation of the circuit under most conditions.

fig **21.7** *master-slave J-K flip-flop logic diagram*

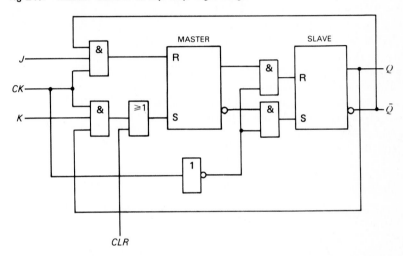

The circuit illustrated is a real device, type SN74107, which has two identical *J-K* master-slave bistables in a single 14-pin package—illustrated in Figure 21.8.

fig **21.8** *a TTL dual master-slave J-K flip-flop, the SN74107*

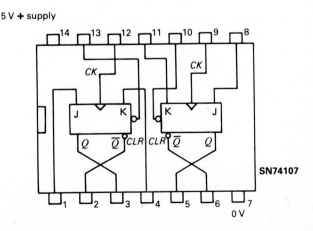

Although the master–slave arrangement appears unnecessarily complicated, it differs from the simpler arrangement in an important respect. If
the clock pulse is at logic 1, a logic 1 applied to J or K will not affect the
outputs. However, new data are accepted by the master. When the clock
pulse returns to logic 0, the master is isolated from the inputs but its
data are transferred to the slave, and Q and $\bar{Q}$ can change state. In a system
incorporating several such flip-flops, this gives the important advantage
that they all change state simultaneously.

21.3 BINARY COUNTERS

It is clear from Figure 21.5 that a single flip-flop actually divides by two:
the waveform from Q (or $\bar{Q}$) is at half the frequency of the clock. The
circuit is actually 'counting' up to two. We can wire the two flip-flops in
the SN7473 together, as shown in Figure 21.9.

fig 21.9 *two flip-flops connected to provide a binary count of 0 to 3*

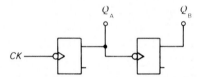

It is understood, when drawing logic diagrams like this one, that terminals not shown are connected to the right logic levels for the circuit to
operate correctly—J, K and CLR are all taken to logic 1, and of course the
circuit's power supply has to be connected. The output of this configuration is illustrated in Figure 21.10. Follow the logic by looking at the two
figures.

fig 21.10 *clock and output waveforms of the two flip-flops shown in
figure 21.9; the table shows the derivation of the binary count*

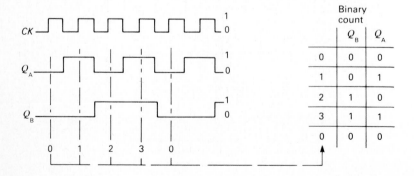

Binary count		
	Q_B	Q_A
0	0	0
1	0	1
2	1	0
3	1	1
0	0	0

This circuit has divided by two twice, or divided the input frequency by four. Also, if we look at the logic levels at Q_A and Q_B we can see that the outputs have provided a count of 0 to 3 in binary. After 3, the counter goes through the cycle again. Each stage doubles the size of the count, so that three stages would count up to 8, and four stages would count up to 16 (including the zero). Four-stage binary counters are available all in one package, such as the SN7493 shown in Figure 21.11.

fig 21.11   *a typical TTL four-stage binary counter, the SN7493*

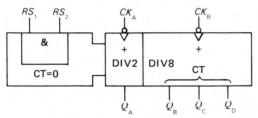

There are two clock inputs in this integrated circuit: CK_A is the normal one used, but CK_B, connected to the *second* flip-flop in the chain, is provided because the Q_A output is not connected to the clock input of the second flip-flop. This gives the user the choice of a three- or four-stage counter; for four-stage operation Q_A is connected externally to CK_B. The two *RS* inputs are used to reset the counter (all outputs to logic 0); both *RS* inputs have to be high to cause reset to take place. These various alternatives are 'designed in' to the integrated circuit to make it as flexible as possible. It makes economic sense for a manufacturer of ICs to aim at the largest possible market!

There are many more types of flip-flop, but perhaps the only other really important one to consider here is the *D-type flip-flop* (*D* stands for data). The *D*-type flip-flop has only one input apart from the clock, and this is connected to the *J* and *K* inputs as shown in Figure 21.12.

fig 21.12   *input connections of the D-type flip-flop*

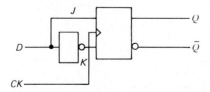

The inverter in the *K* line ensures that the *J* and *K* inputs are always different from each other. A logic 0 or 1 on the data input will flip (or flop) the outputs when the clock pulse is at logic 0. The *D*-type can also be persuaded to toggle, by connecting it as shown in Figure 21.13.

fig 21.13 *connection required to make the D-type flip-flop toggle with the clock pulses*

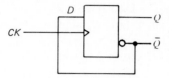

Typical of TTL D-type flip-flops is the SN7474, shown in Figure 21.14, which has two flip-flops in the same package. The 'preset' input is the complement of the 'clear' input, and sets the outputs to the opposite state.

fig 21.14 *a typical TTL dual D-type flip-flop, the SN7474*

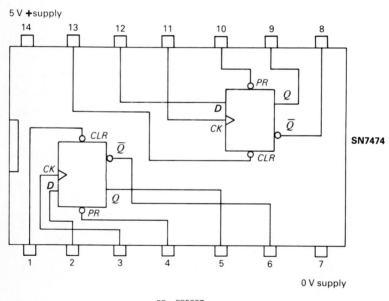

PR = PRESET

Binary counters have many uses, but sometimes it is better to have a counter with a suitable output for ordinary denary (tens) counting. Counters of this kind use ordinary flip-flops for the actual counting, with a system of logic gates to stop the count at nine. Typical of this sort of counter is the SN7490 decade counter. It has four outputs—the output is still in binary—and the sequence of counting is called *binary coded decimal* (BCD) meaning that the count is in binary but stops at nine.

The BCD output can be converted to a useful 'one of ten' output—ten lines, all at logic 0 except one. The one that is at logic 1 steps along the row for each count. To perform the conversion, a circuit called a *BCD to decimal decoder* is used: it has four inputs and ten outputs. A typical TTL circuit of this type is the SN74141. A popular example of a 'one-chip' decade counter—CMOS this time—is the CD4017. This is illustrated in Figure 21.15*a*. The circuit in Figure 21.15*b* makes an interesting demonstration of this type of counter. The LED display can either be made from separate LEDs, or with a 10-segment LED bar display like the one illustrated in Chapter 16 (p. 235). Provided the batter is not more than 9 V, no current limiting resistors are needed, the output current limitation of the CMOS gates being sufficient to ensure safe operation.

Decade counters, along with their associated decoders, can be 'chained' to provide counting and a digital display of the count. Each stage's nine count is used to clock the following stage to provide the carry to the tens, hundreds and thousands counters (and so on, if more stages are used).

Alternative output decoders produce a suitable combination of outputs to drive a 7-segment LED display or LCD.

21.4 LATCHES

It is often useful to 'freeze' the display while allowing the counter to go on counting. This may be essential to check the total of a fast counter

fig 21.15

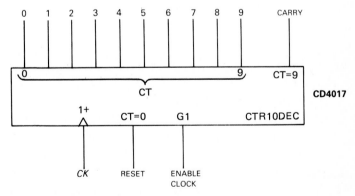

RESET: Normally at logic 0; when at logic 1, this resets the counter (zero output high) and prevents the counter operating

ENABLE CLOCK: Normally at logic 0; if at logic 1, clock pulses have no effect on count

CARRY: Long pulse, goes from logic 0 to logic 1 as count goes from 9 to 0; used instead of 9 output for 'chaining' counters

(a) *a CMOS decade (one-of-ten) counter*

312

fig 21.15 cont.

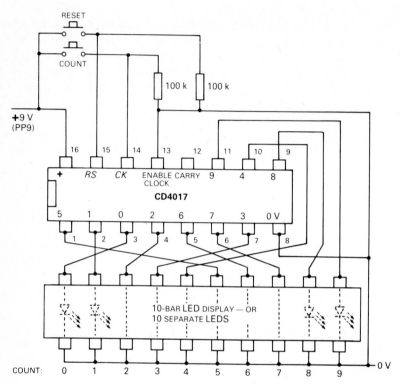

(b) *circuit for demonstrating the decade counter; only one of the LED segments is lit at any time, the segment corresponding to the count; pressing RESET sets the count to zero, and each successive press of the COUNT button advances the count by one; if the switch does not operate 'cleanly', contact bounce may cause the count to advance several steps (try using different switches); the supply voltage of 9 V should not be exceeded with this circuit, since there are no external limiting resistors for the LEDs*

without interrupting the count. A circuit known as a *latch* is needed for this. A latch has four inputs and four outputs. Normally the four outputs assume the same state as the inputs, but when a control input (known as the *strobe* input) is taken to logic 1, the outputs are locked in whatever logic state they were in at the instant of the strobe input going high. This enables the display to be frozen, without affecting the counter(s). A TTL circuit of this type is the SN7475 *quad bistable latch*.

As you might expect, it is possible to combine the BCD counter, latch and decoder on a single IC, which saves wiring and simplifies the construction of a complete counter system. The SN74143 illustrated in Figure 21.16 is a good example of this type of integrated circuit.

fig 21.16 *a TTL BCD counter/latch/decoder circuit, the SN74143*

SN74143

21.5 SYNCHRONOUS COUNTERS

The simple binary counter illustrated in Figure 21.9 is usually described as an asynchronous or 'ripple-through' counter. The reason is that it takes a certain amount of time for each stage of the counter to operate, and the more stages there are, the longer it is before the clock pulse has propagated through the system and all the outputs have settled down to their steady-state values. There are various problems associated with this mode of operation. If you are designing a high-speed system, it is essential to know exactly when each switching takes place, or else logic sequences can get out of step; this can cause instability which prevents an apparently sound design from working.

A counter that works in a more predictable way is the *synchronous counter*. Synchronous counters have the counting sequence controlled by a clock pulse, all the outputs of the flip-flops changing state simultaneously (or synchronously). Synchronous counters are quite complicated, and involve gating the inputs of an array of J–K master–slave flip-flops. Outwardly its operation looks the same as an asynchronous counter, apart from the fact that *all* changes in the outputs take place as the clock pulse falls to logic 0.

21.6 UP-DOWN COUNTERS

Counters are also available to count up (add to the count) or count down (subtract from the count). Such counters feature an UP/DOWN control input which determines which way the counter works, according to the logic state applied to it.

21.7 CMOS COUNTERS

Because most of the circuits so far considered are from the TTL family, you might think that CMOS counters are less common. This is not the case, and there are CMOS counterparts (with slight variations) for all the types discussed already. For medium- and low-speed operation—up to 5 MHz—the advantages of CMOS counters are considerable. The low power requirements, wide power-supply voltage tolerance and excellent noise immunity make CMOS the first choice for low-speed/low-cost applications—and incidentally for easy and reliable demonstrations!

21.8 SHIFT REGISTERS

A shift register is in many ways similar to a multi-stage counter, but it is used to store binary data and to 'shift' the data to the right or left along a row of output lines. Shift registers form an important part of all computers. Figure 21.17 shows a typical four-stage shift register (CMOS)— the CD4015 contains two identical registers.

This type of shift register is known as a *serial-in, parallel-out* design, and it works like this. When the clock pulse (CK) goes to logic 1, the logic level currently on the DATA input is transferred to Q_1. The next clock pulse also 'reads' the data input and transfers it to Q_1, and the logic level previously present on Q_1 is shifted to Q_2. The third clock pulse reads the data input again, sets Q_1 to the corresponding logic level, and shifts the old data on Q_1 to Q_2, and that on Q_2 to Q_3. Each new clock pulse shifts the data one place to the right. At the end of the register (Q_4 in this case) data are lost. It is possible to connect shift registers together to add stages—progress of data through an eight-stage shift register (such as could be made using the two halves of the CD4015) is charted in Figure 21.18. The RS (reset) input allows the register to be reset, making all the outputs logic 0 (see Figure 21.17).

fig 21.17 *a CMOS dual four-stage shift register, the CD4015*

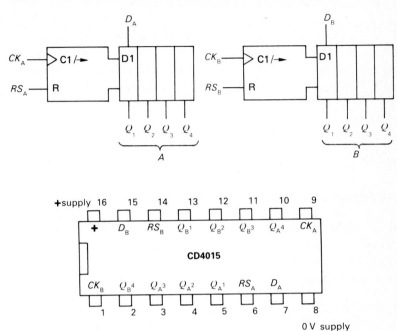

fig 21.18 *the passage of data through an eight-stage shift register*

There are many variations in shift register types. Some are *serial-in, serial-out,* and simply accept an input, which appears at the output after a fixed number of clock cycles—they are like the one shown in Figure 21.16 but with only the Q_4 output. Others have a *parallel-in* facility, by means of which data can be loaded into the register by applying them to a number of inputs (one for each stage) simultaneously. Still others are *bidirectional* and can shift data either way according to the logic level applied to a control input.

21.9 INTEGRATED CIRCUITS FOR COUNTING

Finally, it must be made clear that there are a large number of different kinds of ICs for counting in both major logic families, and many special-purpose counting ICs using PMOS or NMOS techniques. Most are variations on the basic type of counters, shift registers, latches and decoders described in this chapter, and many are combinations of these circuits. One thing is certain. These days it is very unlikely that anyone will have to design a counter circuit using discrete components. The current range of ICs is so large, and the individual devices so cheap, that 'doing it the hard way' would hardly ever be justified. When selecting an IC—particularly a counter or shift register—it is important to read the manufacturer's data-sheet carefully. All relevant specifications and parameters should be listed in great detail. Systems that refuse to work are often the result of mistakes in reading the data-sheet, or of not bothering to read the data-sheet at all.

QUESTIONS

1. Give another name for the *flip-flop*.

2. An electronic stop-watch would probably incorporate latch circuits. Describe the function of a latch.

3. The following data are applied to an 8-bit shift register:

$$0 \quad 0 \quad 1 \quad 1 \quad 0 \quad 1 \quad 0 \quad 1$$

What data would be present after three clock pulses, assuming the input remains at logic 0?

4. Sketch a system diagram for a TTL 0 to 9 counter, driving a 7-segment LED display.

5. What are the functions of the CLEAR and SET lines of a counter IC?

DIGITAL SYSTEMS – TIMERS AND PULSE CIRCUITS

Every counting circuit has to start with a circuit to generate pulses to count. Rarely, the pulses may be produced by some source outside the circuitry, for instance when counting events. More usually, the pulses are generated by special pulse-generator or timing circuits.

This chapter begins by looking at various methods by which pulses suitable for digital systems can be produced, and then goes on to consider the design of a 'real-life' digital system, as an example of the way digital systems are put together from gates and other commonly available ICs.

22.1 TIMERS

Although timers can be developed from discrete components, it is usual to design this part of the system using ICs. The most commonly used timer circuit is the NE555, packaged in an 8-pin DIL pack and sold for the price of a cup of coffee. The NE555 is quite an elderly design, and has a more modern variant, the ICM7555 of 'low-power 555'. The latter uses CMOS technology instead of bipolar, and operates with about one-hundredth of the supply current; otherwise, the two ICs are identical. The NE555 can be replaced directly with the ICM7555, but the reverse is not necessarily true. The package and pin connections are given in Figure 22.1.

The NE55/ICM7555 ('55' for short in this chapter, and often in the industry) is very versatile, and can be used as a *monostable* or as an *astable*. Pins 1 and 8 are the power supplies. The functions of the other pins become clearer by reference to a diagram of the system inside the 555. This is illustrated in Figure 22.2.

The basic monostable configuration is given in Figure 22.3. When the *trigger* is taken from +5 V to 0 V (logic 1 to logic 0) the output will go high for a period determined by the values of R and C. The length of the output pulse is given as 1.1 RC seconds, so in the circuit of Figure 22.3

fig 22.1 *a popular timer IC, the NE555*

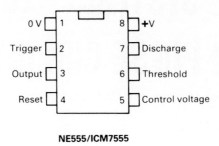

NE555/ICM7555

fig 22.2 *system diagram of the internal circuitry of the NE555*

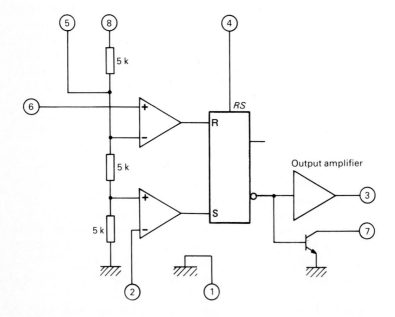

will be $1.1 \times (220 \times 1000) \times (100/1\,000\,000\,000)$, which is 0.0242, or about 24 ms. The calculation can be made easier to remember if you use kΩ and μF, and this provides an answer in milliseconds directly when used in the formula in place of R and C.

In this circuit the timing sequence will continue even if the trigger is taken back to logic 1. To stop the timer, the *reset* input is taken from logic 1 to logic 0.

fig 22.3 *the NE555 in a monostable configuration*

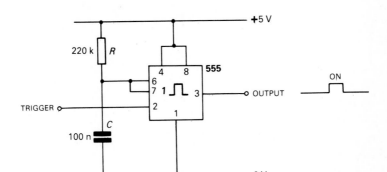

Because the 555 can deliver currents of 100 mA or more, it is possible to use the 555 to drive a relay, for example, directly. However, precautions need to be taken to ensure that the back e.m.f. generated by the relay coil when the relay is switched off does not cause damage to the circuits. A diode placed across the coil in the 'opposite' direction to the current flow will absorb the induced high voltage. Where relays are packaged for use in digital circuits (for example, DIL reed relays) a diode is often connected across the coil inside the encapsulation.

The output of the 555 is, however, ideal for driving TTL or CMOS logic ICs, as it has a very rapid rise and fall, required for accurate timing and correct operation of some systems.

The 555 can be used in an astable mode, to generate a continuous train of pulses. The 'on' and 'off' times can be controlled separately, within certain limits. Figure 22.4 shows a typical astable configuration.

fig 22.4 *the NE555 in an astable configuration*

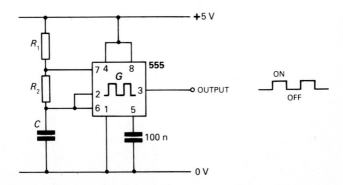

Note that the 100 nF capacitor connected to pin 5 is required for the NE555, but is not needed for the CMOS version—by reference to Figure 22.2, can you see why?

The 'on' time (logic 1) is given by:

$$0.693\,(R_1 + R_2)\,C$$

and the 'off' by:

$$0.693\,(R_2)\,C$$

so it is evident that the 'on' time must always exceed the 'off' time in this configuration, unless R_1 has a very low resistance compared with R_2, in which case they can be about equal. The output sense can always be reversed (1 for 0) by the simple step of following the 555 with a TTL or CMOS inverter.

The 555 is very accurate, and is affected very little by temperature and voltage variations. The ICM7555 can be used with any supply voltage in the range 2-18 V, and timing accuracy changes by only about 50 parts per million per degree C. Maximum operating frequency is 500 kHz, and the minimum frequency is determined only by the leakage of the capacitor used; one cycle per several hours is achievable.

Both TTL and CMOS logic gates can be used as timers, provided the frequency required is not too low. A simple CMOS monostable is shown in Figure 22.5. The input to the inverter is normally held at logic 0 by resistor R. If the input is rapidly taken to logic 1, the input of the gate goes to logic 1 as well, and the output goes to logic 0. However, the capacitor now charges via R, since the left-hand side (in the diagram) is at logic 1 level. The charging rate is governed by R (assuming R to be at least a few kilohms), and the voltage across the capacitor drops at a controlled rate. When the voltage on the input of the gate drops to half the supply voltage, the gate changes state and the output goes high again.

fig 22.5 *a simple CMOS monostable*

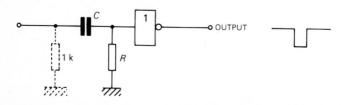

The shaping of the input and the requirement for an input resistor (1 kΩ in the diagram) to discharge the capacitor are taken care of if the circuit is preceded by another gate (see Figure 22.6). With the input to the first gate at logic 1, the output is also a logic 1. If the input is taken to logic 0, the output is a logic 0 pulse, of duration $0.5\,RC$. Note that the input must be held at logic 0 throughout the duration of the output pulse. There will be no further output from the circuit until the input is taken back to logic 1 and then to logic 0 again.

fig 22.6   *the CMOS monostable is usually preceded by a CMOS gate, which takes care of the correct input parameters*

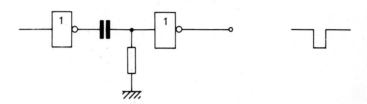

CMOS circuits may also be used to make astables. Figure 22.7 illustrates the standard CMOS astable configuration. The time taken for one complete cycle of this astable circuit is approximately $1.4\,RC$, with a small variation according to supply voltage. The circuit is remarkably unaffected by temperature variations, and no temperature compensation will be required.

fig 22.7   *a basic CMOS astable circuit*

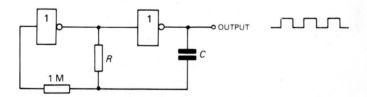

It is possible to control the 'on and 'off' times independently by forcing the capacitor to charge and discharge through different resistors—easily done with diodes. The circuit of Figure 22.8 provides an output pulse that can be varied from 1 to 2 ms, with a fixed interval between pulses of about 20 ms.

It is possible to make similar circuits using TTL logic, though for timing applications a gate with a Schmitt input should be used. Normal TTL responds erratically to input signals that rise and fall slowly; the Schmitt input range of gates provides a rapid switching transition even when the input is a 'slow' rise or fall. A TTL astable design is shown in Figure 22.9.

Although CMOS and TTL astables can be quite accurate, and some of the purpose-made timers are very accurate, their accuracy depends even-

fig 22.8 *a CMOS astable circuit that has independently variable 'on' and 'off' times*

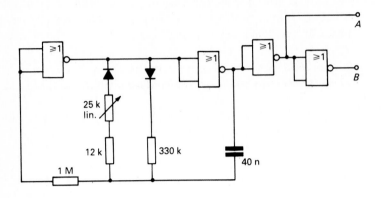

fig **22.9** *a TTL astable configuration*

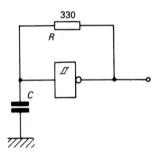

tually on the performance of the resistor and capacitor used as timing elements. For some applications, such as clocks and watches or computer clock generators, this is not accurate enough. As we saw in Chapter 10, the most accurate form of oscillator in general use is the quartz crystal oscillator. A crystal-controlled oscillator can be made using a CMOS gate, as in the circuit of Figure 22.10 overleaf. The crystal should not have an operating frequency of more than 3 MHz. The variable capacitor can be used for trimming the frequency within narrow limits.

22.2 IMPLEMENTATION OF A DIGITAL SYSTEM

Traffic lights in Britain go through a sequence of green (*go*), amber (*stop if you can*), red (*stop*), red and amber (*on your marks* . . .), and back to green again. The design of a timing system to operate such traffic lights,

fig 22.10 *a crystal oscillator circuit using a single CMOS gate*

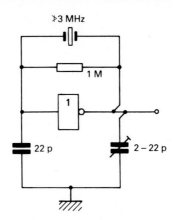

even without radar control, car counters and other 'extras', turns out to be by no means trivial.

First, assuming that we have excluded systems of motors, cams and switches, we must choose the kind of technology. CMOS is probably the best, not because we are worried about power requirements, but because CMOS is less affected by electrical interference than is TTL. Also, CMOS happens to be a better bet for prototype demonstrations! Temperature range, -55 to $125\,^{\circ}$C, should be satisfactory for Britain. It is essential to design the power supply to protect the circuits from transients in the power line (suppose the power-supply lines are struck by lightning?) and it is possible that this consideration alone could tip the scales in favour of TTL.

Assume that we settle on CMOS.

There are four possible configurations for the lights, shown in the table in Figure 22.11.

fig 22.11 *truth table for UK traffic lights*

1	2	3	4	
		R	R	Red
	A		A	Amber
G				Green

Three gates will 'translate' the four outputs into the right combinations of lights—two NOR gates and an inverter (see Figure 22.12).

Each of the four lines must go high in turn, so a counting circuit would almost certainly be the best bet—a CMOS CD4017 (see Figure 21.15 on

fig 22.12   *a system of gates to implement the truth tables in Figure 22.11*

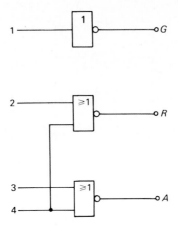

pp. 306–7) seems ideal, except that there are too many output lines. The counter must be driven by timer, to 'clock' the count. The circuit given in Figure 22.13 works.

Note the rather smart connection between output 4 and the RESET input: when the clock sends the CD4017 to a count of 5, output 4 goes high and makes RESET high as well, forcing the counter to reset back to zero, with line 0 high. This works with the CD4017, but might not work with other counters, which need a longer pulse on RESET. In these cases a monostable needs to be fitted between the output and the RESET input. It all depends on the internal organisation of the IC.

fig 22.13   *driving the gates with a CD4017 counter*

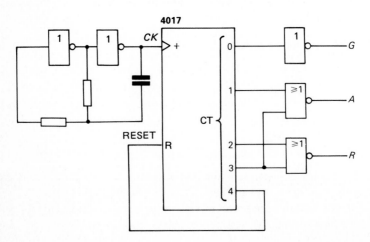

The 555 generates clock pulses, to time the lights.

Drivers will have spotted the snag with the design—the lights need to be red or green much longer than they are amber or red and amber. Figure 22.14 shows a sort of 'stop-gap' design, in which the lights are green or red for four times as long as they are amber or red and amber. It does this by making use of all ten output lines of the CD4017, plus larger NOR gates to decode the outputs. Operation is straightforward and should need no explanation beyond the circuit diagram.

fig 22.14 *using the counter to provide more realistic timings for the lights*

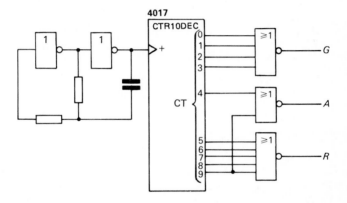

What is really needed is a system in which the timer's frequency is altered for each light—or perhaps separate timers for each of the four possible intervals. It turns out that the second idea is quite practical and relatively simple to implement. Figure 22.15 illustrates a complete system. This one *does* need explanation.

Assume that the circuit initially has output 0 high. The upper input line of the 4-input NOR gate (G_1) is at logic 1, so the output of the gate G_1 is at logic 0. After a period determined by the RC timing network ($100\,\mu$F and $1\,$MΩ) the input to G_1 will drop to a point where it goes from logic 1 to logic 0. With all its inputs now at logic 0, G_1 output goes high, applying a logic 1 to G_2 input via the 4.7 nF capacitor; G_2 output goes to logic 0 until the time constant $0.5\,RC$ has elapsed ($R = 120\,$kΩ, $C = 4.7$ nF, gives around 250 μs). At this point G_2 output goes high again, and the leading edge of this pulse clocks the 4017 counter. Output 1 goes to logic 0, and now output 1 becomes logic 1.

The whole sequence is repeated, but with a different timing network, 4.7 μF and 1 MΩ. The counter steps to outputs 2 and 3, and finally to 4, which forces a RESET and goes back to output 0.

fig 22.15 *a practical demonstration circuit giving independently variable 'on' times for red and green traffic lights*

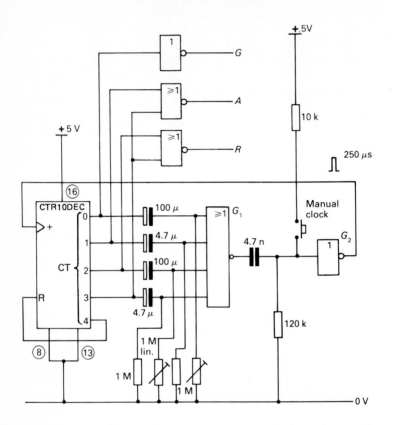

Three gates decode the outputs to operate three lights, as in the earlier system. In this design the amber and red and amber time constants are fixed at about 2.5 s. The green and red can be separately set up to approximately 50 s.

As a working system this might or might not prove practical when fitted into real traffic lights. There is one design difficulty to be sorted out, too. If (after a power cut) the counter powers up with outputs 5, 6, 7, 8 or 9 high, there is no reason why the circuit should ever start. A manual clocking switch solves the problem for the experimental model, however, an automatic start-up would be needed in practice.

There is also the question of making the CMOS outputs drive three heavy-duty light bulbs. This would probably be done with triacs, and perhaps opto-isolators. Less impressively, LEDs can be used for demonstration purposes—see Figure 22.16.

fig 22.16 *a traffic light indicate circuit for use with the circuit in Figure 22.15 for demonstration purposes*

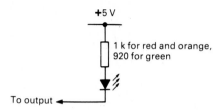

The whole system is working near the upper limit of timing by means of CMOS gates; the leakage of the timing capacitors (which should be low-leakage types such as tantalum beads) is becoming too significant. In a real-life traffic light, the electronic system would account for only a very small part of the cost. Maintenance costs are high, and a results of a failed traffic light chaotic, so high-reliability components (555 timers, at least) would be used.

But the exercise of designing logic systems is the same, and the designer must balance many different, and often conflicting, requirements to arrive at an optimum design.

22.3 PULSE-WIDTH MODULATION AND PULSE-POSITION MODULATION

Timing circuits can be used to encode information, either for transmission down wires or fibre-optic systems, or for modulating radio frequency carriers. There are two possible ways of sending information. If extreme accuracy is needed, then a purely digital method can be used, sending a stream of numbers that represent instantaneous values (as in digital sound recording). This is relatively complex and expensive. Where less accuracy is called for, a method known as *pulse-width modulation* can be used. Pulse-width modulation (PWM) works by transmitting a stream of pulses, the duration of each pulse corresponding to a varying quantity. The principle is demonstrated in Figure 22.17, which shows a sequence of pulses representing an increasing value. The *resolution* of the system will depend upon how finely the timing of individual pulses can be timed at the transmitter and measured at the reciever. PWM techniques are often used for transmission of digital data from computers, a long pulse representing a 1 and a short pulse representing a 0. The resolution of such a system need only be sufficiently good to distinguish between two possible values!

A modification of PWM is *pulse-position modulation* (PPM), which uses fixed-length pulses but varies the time between them to represent

fig 22.17 *pulse-width modulation*

Pulse width represents value

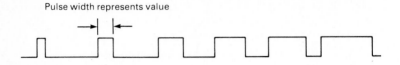

fig 22.18 *pulse-position modulation*

Time between pulses represents value

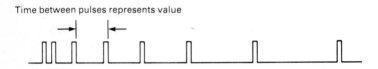

analogue values. Figure 22.18 illustrates a PPM sequence, again representing a steadily increasing value.

With reference to the traffic-light timing system in Figure 22.15, it should be clear how such pulses can be generated using digital circuits and timers. If the system does not need to respond quickly, relative to the pulse frequency used, both PWM and PPM can be *multiplexed* to provide information about several different variables. A counter in the transmitting system 'looks at' different variables in turn, and a similar counter in the receiver uses the interval between relevant pairs of pulses to recreate the original data. The traffic-light controller system of Figure 22.15 actually produces a PPM signal which is applied to the clock input of the 4017. The 'events' that are represented by the spaces between the 250 μs clock pulses are the times that each light is on. Figure 22.19 shows the waveform for one complete cycle of the traffic lights, assuming that the 'green' is about 35 seconds and the 'red' is 40 seconds.

This train of pulses contains all the necessary information to operate traffic lights at some point remote from the timers. As with the television

fig 22.19 *the waveform for one complete cycle of the traffic lights: pulse-position modulation*

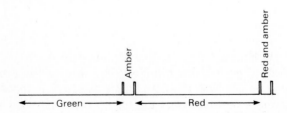

transmission, it is necessary to devise a system to synchronise the receiving counter with the transmitting counter; this might be a long pulse, a short burst of closely spaced pulses, or even a temporary cessation of the transmission. Synchronisation is necessary to ensure that the multiplexing of the receiver is looking at the same channel as the transmitter.

Figure 22.20 illustrates a PPM coded sequence for eight channels, with a long pause used to synchronise the transmission. The first four channels are showing a lower value than the last four. Each channel is 'updated' once every 20 ms or so, and it is not possible to change information at the receiving station more rapidly than this. The resolution of each channel is unaffected by the rate at which the signal is multiplexed, depending only on the accuracy of timing circuits in the transmitter and receiver.

fig 22.20   *pulse-position modulation for a multi-channel control transmission; the long gap is used to synchronise the counters at the transmitter and receiver*

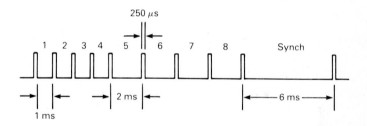

At the receiver, the signal can be decoded by a simple counter, with added circuitry for the synchronising system. A suitable circuit is given in Figure 22.21; this would decode the waveform of Figure 22.30 into separate pulses.

The clocking signal is applied to the input of a 4017 counter, and also to the RESET input. RESET is normally held positive by the resistor, the capacitor being fully charged. When the input waveform goes negative, the capacitor is discharged through the diode and RESET goes low, allowing the counter to work. The time constant of the circuit is chosen so that RESET rises above half the supply voltage only during the 6 ms synchronising pulse.

During the gap in the pulse train used to synchronise the system, the capacitor discharges beyond the point where the potential on the RESET input rises above half the supply voltage, causing the counter to reset to zero. Note that the receiver will operate correctly with any number of channels, from one to ten, provided the synchronising system is working.

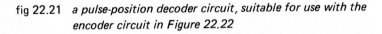

fig 22.21   *a pulse-position decoder circuit, suitable for use with the*
encoder circuit in Figure 22.22

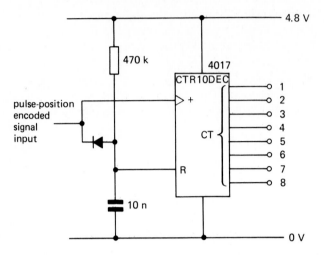

If it appears that the principles of PPM transmission and reception have been given in greater detail than might be expected, then it is for a reason. The system described above is the standard method used for the radio control of models, and we can use the principles explained in this chapter, in conjunction with the transmitter and receiver circuits described earlier, to produce a working model control transmit/receive system.

The circuit in Figure 22.22 represents the complete encoding and modulation system for the transmitter illustrated in Figure 15.3 (p. 199) and 15.13 (p. 207). In essence it is clearly very similar to the traffic-light controller in Figure 22.15, and the principles of operation are much the same. A second gate (G_3) inverts the output, and a single *pnp* transistor is used to increase the output current capability (a transistor used in this way is referred to as a 'buffer').

The circuit uses a 4022, an IC that is almost identical to the 4017, but counts up to eight instead of ten. The transmitter encoder provides seven identical independent channels of control. The synchronising interval is generated by $R_s C_s$, providing a fixed pulse of about 6 ms. All the other channels are variable between 1 and 2 ms, the standard values that represent 'full left' and 'full right' to a model control servo (see below). In the centre position, the pulse interval is 1.5 ms ('straight ahead'). Supply decoupling (see Chapter 10) is provided by the 220 Ω resistor and the 10 μF capacitor, which prevent transients on the power-supply lines from triggering the 4022 spuriously.

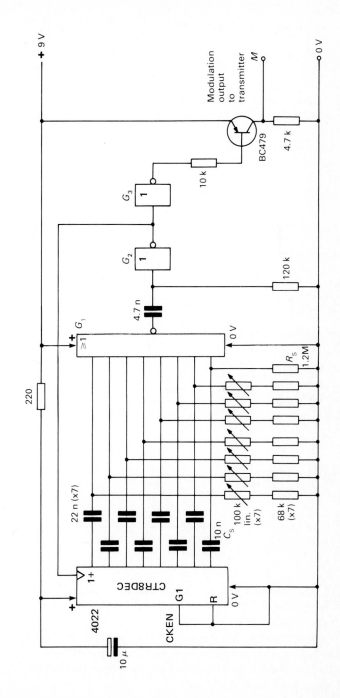

fig **22.22** complete coding system for an 8-channel radio control; the output of this circuit can be used to modulate the transmitter described in Chapter 15

The radio control receiver decoder works on this principle, the input to the decoder being driven by the receiver's output. CMOS circuits are relatively unaffected by voltage fluctuations resulting from heavy loads imposed on the battery by servos. Not so the receiver circuits. In order to hold the supply voltage to the receiver constant, decoupling is required in the power line between the decoder and the receiver. A simple decoupling circuit (Figure 10.12) is not effective enough, so a transistor is used to improve the performance. This circuit is similar to the simple regulator circuit shown in Figure 10.26, but with a capacitor replacing the Zener diode.

fig 22.23   *the complete radio control receiver decoder, including a voltage regulator for supplying the receiver. This circuit can be used with the receiver described in Chapter 15*

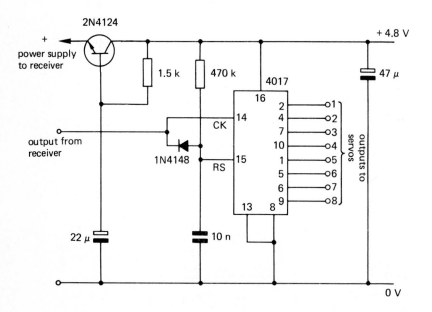

The decoder and decoupling circuit is shown in Figure 22.23. This circuit connects directly to the circuit in Figure 15.23 to make the complete radio control receiver.

Model radio control servos are best obtained ready-made, as the mechanical parts are small and have to be precise. The electronics in the servo convert the 1–2 ms pulse into a mechanical movement, the system diagram being given in Figure 22.24.

fig 22.24 *system diagram for a typical model control servo*

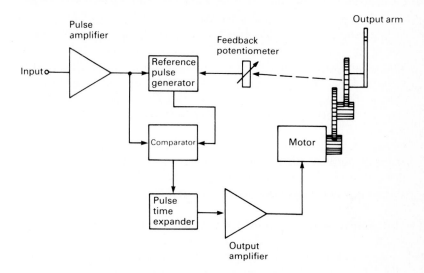

A specially designed IC is used, the NE544. This, along with a few discrete components, a motor, gears and feedback potentiometer, are all squeezed into a plastic case about 40 × 30 × 20 mm!

Model servos are designed to be used with multi-channel systems, and accept an input pulse supplied at the rate of one every 20 ms or so. Most servos require a 'positive pulse', which is the type used for the receiver described in this book.

Servos can be tested with a circuit that provides the requisite pulses—the circuit of Figure 22.8 is ideal! The two outputs provide suitable pulses for either type of servo.

The operation of the servo is typical of a *feedback control system*, and works as follows. A small, high-efficiency motor is connected through a reduction gear system to the servo output arm, and also to a potentiometer. This potentiometer is connected to the reference pulse generator (see Figure 22.24). Imagine the servo is in its 'centre' position when a 1.5 ms pulse is received. After amplification the pulse is fed to the comparator. The incoming pulse also triggers the reference pulse generator, which, while the potentiometer is in the middle of its travel, is also 1.5 ms long. While the pulses are the same length, there is no output from the comparator.

If the incoming signal is now reduced to 1 ms (command for 'full left'), the comparator detects a difference between the incoming pulse and the reference pulse—the incoming pulse is 0.5 ms shorter. This causes an output from the comparator, which, via the pulse time expander and output amplifier, drives the motor. The motor rotates both the output arm and the control potentiometer until the reference pulse is also 1 ms long and the output from the comparator ceases.

Increasing the input pulse length to 2 ms ('full right') again causes an output from the comparator, but because the incoming pulse is now longer than the reference pulse—by 1 ms—the output is of the opposite polarity. Once again, power is applied to the motor, but in the opposite sense so that the motor turns the other way. The output arm and potentiometer again turn until the reference pulse once again matches the input pulse.

Advantages of this system are (i) the servo uses little power unless it is changing position, (ii) the motor will work against any attempt to force the output arm into a different position, and (iii) a lot of torque is available, better than 2 kg/cm for an average servo.

This chapter, dealing principally with digital circuits and timers, has actually brought together several aspects of electronics and electronic systems.

Timers have led to the design and development of a traffic-light control system. Digital transmission of data, based on timers, has led to a data-encoding and decoding system. And finally, the data system has been used to modulate a transmission that can be received by a receiver system (both transmitter and receiver having been described elsewhere in the book). The result is a practical system, fully compatible with current commercial model control systems, that can form the basis of an interesting project for students.

When using the radio-control system, be careful about safety. The 27 MHz band, used for model boats, cars and aircraft, is open to interference from CB radios working at 27 MHz on different but very close frequencies. Power model aircraft can be dangerous, and the 35 MHz FM band, also available in the UK for aircraft only, is preferable in this case.

In the next chapter, we shall examine systems that are 'pure' digital electronics, operating without the use of 'analogue' timers.

22.4 DIGITAL TO ANALOGUE CONVERSION

Although the circuits we have been considering have transmitted information by means of digital pulses, the information content has actually been in an analogue form. The time between each pulse is a continuously variable quantity and is therefore analogue in nature, and not digital. However, it is often convenient to treat analogue quantities (sounds, movement, heat, light) as numbers for the purpose of digital processing. A typical example is the digital recording system outlined at the beginning of

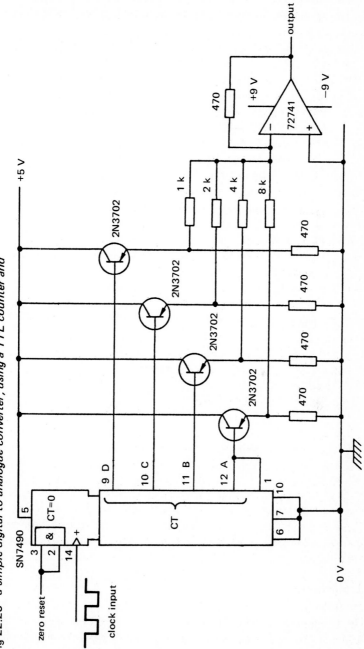

fig 22.25 a simple digital to analogue converter, using a TTL counter and

Chapter 18. Before such a system can be developed, we need systems for converting digital information to analogue values, and vice versa. Several techniques exist.

It is relatively easy to translate a binary number into a voltage. The circuit in Figure 22.25 is typical. The SN7490 4-bit binary counter drives four 2N3702 *pnp* transistors, one for each bit, that are either cut off or fully on according to whether the bit is zero or one. A proportion of the emitter voltage of each transistor is applied to the inverting input of an SN72741 operational amplifier. The voltage from each transistor is proportional to the value of the bit, each successive higher bit applying twice the voltage of the previous bit. The op-amp sums the voltages, and the output is a voltage that is an exact analogue of the binary number. For obvious reasons, this kind of *digital to analogue converter* (abbreviated to DAC) is called a *weighted resistor DAC*. Figure 22.26 shows the output for an increasing count.

fig 22.26   *the output of the converter in figure 22.25, as the count increases*

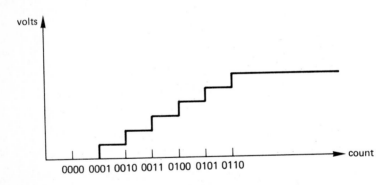

To all intents and purposes the weighted resistor DAC operates instantly. The analogue output follows the digital input without any delay other than those inherent in the semiconductor devices being used. The opposite kind of conversion, *analogue to digital* (ADC) almost always takes time, and different circuits are used according to the design requirements.

22.5 ANALOGUE TO DIGITAL CONVERSION

The simplest kind of ADC is the *single-slope conversion ADC*. The system is shown in Figure 22.27. The principle is quite easy to understand. The counter is reset and the voltage to be converted is applied to the inverting input of an op-amp. The non-inverting input is held at zero volts, as the output of the DAC (the circuit in Figure 22.25) is zero because the count is zero. The output of the op-amp is therefore positive. This applies a 1 to the AND gate. The clock generator puts a square-wave on the other input of the AND gate, and the output of the AND gate turns on and off with the clock waveform. This makes the counter count up, and the output from the DAC rises in a series of steps. There comes a point when the output voltage from the DAC exceeds the input voltage being measured. When this happens the output of the op-amp swings from positive to negative, applying a logic 0 to the AND gate and cutting off the clock signal to the counter. The count stops increasing, and the digital equivalent of the input signal can now be read on the digital output lines of the DAC. The state of the output of the op-amp is used to indicate when the digital output is ready to be read. Once the value has been read by other circuits – and perhaps latched into a display – a pulse can be applied to the RESET input of the counter. The cycle then repeats.

fig 22.27 *schematic diagram of a single-slope conversion ADC*

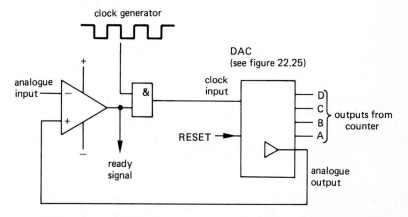

It is clear that although it will provide a continual digital conversion, this kind of circuit takes a little while to operate. The length of time taken for a 'reading' is actually proportional to the voltage being converted and the speed of the clock. If the circuit is intended to provide a conversion that follows, or tracks, the input voltage, then it can be modified to improve the performance. The *tracking ADC* (or servo-type ADC) uses a similar principle except that the counter is an IC designed to count up or down. It is connected in such a way that it counts up if the op-amp used as a comparator shows that the input voltage is higher than the last value, and down if it is lower. This increases the speed of each conversion as the counter only has to count up or down by an amount corresponding to the difference between the current input voltage and the last one.

Figure 22.28 illustrates a different type of single-slope ADC. A capacitor is charged up at a known rate from a known voltage source, so it is possible to predict exactly what voltage it will have reached after a set time. The input voltage (the voltage to be converted) is applied to one input of an op-amp, and the voltage across the capacitor is applied to the other. A stable clock-pulse is gated by the output of the op-amp. Conversion takes place as follows.

fig 22.28 *schematic diagram of a single-slope conversion ADC, using a different technique from the one employed in figure 22.27*

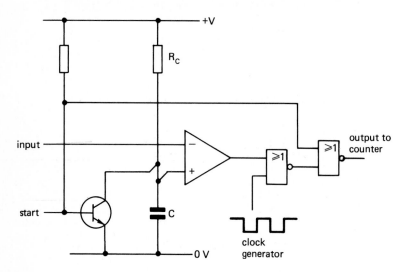

The capacitor, initially uncharged and holding zero volts on the op-amp's input, begins to charge up. Clock pulses appear at the output. When the voltage across the capacitor reaches a level where it just exceeds the input voltage, the output of the op-amp changes polarity, and the clock pulses no longer appear at the output. The number of clock pulses that have been delivered to the output since conversion started depend on the time taken before the voltage across the capacitor and the input voltage are equal. Thus the number of clock pulses delivered to the output is proportional to the input voltage. The digital equivalent can be found simply by counting the pulses.

The above circuits all use single-slope conversion, in which a steadily increasing voltage is compared with the unknown input voltage. For many applications greater accuracy can be obtained using *dual-slope conversion*, shown in Figure 22.29. The capacitor is charged up from the input for a fixed period, equal to the length of time it takes for the counter to reach its maximum count. When this happens, a 'carry' output from the counter disconnects the input and discharges the capacitor through a known resistance. The counter is incremented by the clock pulses until the capacitor has discharged, whereupon the counter shows the digital equivalent of the input voltage. A graph of the voltage across the capacitor shows a steady increase while it charges, then a steady decrease while it discharges – the 'dual-slope' in the name of the circuit. Dual-slope ADCs are slow, but are often used for meters because they are relatively immune to input voltage transients that can make faster systems give erratic results.

fig 22.29   *schematic diagram for a dual-slope conversion ADC*

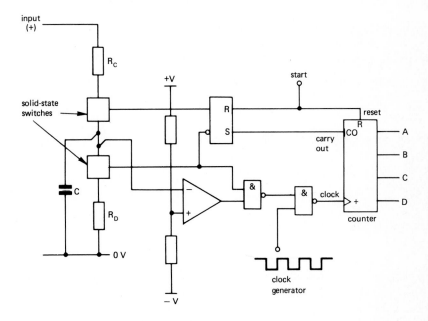

One of the fastest types of ADC is the *successive approximation ADC*. In this design, a voltage corresponding to the value of each bit is compared with the input voltage. The bit is then either set or not set, and, if set, the equivalent voltage is subtracted from the input voltage. This system is complicated, but requires only one clock cycle per bit to carry out the conversion.

The very fastest conversion, such as might be used for audio work, is carried out by using a separate op-amp for each bit, comparing the input voltage simultaneously with voltages that will result in a positive (1) or negative (0) output from the op-amps corresponding to each bit. This kind of circuit operates at a speed depending only on the slew rate of the op-amps, and is called a *clockless ADC.*

All the different types of DAC and ADC are available in integrated circuit form, and there are many different designs, all slightly more suit-able for one or other application.

QUESTIONS

1. Sketch a system diagram in which the output of a timer operating in astable mode is turned on or off by a logic gate.

2. Sketch the circuit diagrams for CMOS and TTL astable circuits.

3. Draw a graph showing the pulse sequence 0111001 as a pulse-width modulated waveform.

4. Draw a graph showing a triangular waveform. Draw a graph of the same waveform, on the same time scale, as a pulse-width modulated transmission.

5. At what point on the excursion of the input from logic 0 to logic 1 does the output of a CMOS gate change state?

DIGITAL SYSTEMS –
ARITHMETIC AND MEMORY
CIRCUITS

23.1 ARITHMETIC CIRCUITS

We are all familiar with the rules of simple addition regarding the 'carry'. Here is an example of an addition sum an 8-year-old might do at school:

$$
\begin{array}{r}
1\ 9\ 7\ 7\ 5 \\
7\ 4\ 2 \\
\end{array} +
$$

$$
\begin{array}{r}
2\ 0\ 5\ 1\ 7 \\
1\ \ 1\ \ 1 \\
\end{array}
$$

The child puts in the 'carries' below the answer line (so do I!) as a reminder that there is a figure to be carried from the previous column. Actually, even the child is doing the calculation in short-hand. The long way is to do the sum as two *half-additions*, like this:

$$
\begin{array}{r}
1\ 9\ 7\ 7\ 5 \\
7\ 4\ 2 \\
\end{array} +
$$

1 9 4 1 7	PARTIAL SUM NUMBER ONE
1 1	CARRIES FROM FIRST HALF-ADDITION
1 0 5 1 7	PARTIAL SUM NUMBER TWO
1	CARRIES FROM SECOND HALF-ADDITION
2 0 5 1 7	SUM

The operation to add the first lot of carries is performed to provide the second partial sum; but this produces its own 'carries', which are added in to provide the answer. The same principle, that of using two half-addition operations, can work equally well with binary arithmetic, and is the method used in digital systems to add two binary numbers together.

Here is the calculation to add two binary numbers, the binary equivalents of (denary) 103 and 18:

$$1\ 1\ 0\ 0\ 1\ 1\ 1 \\ \underline{1\ 0\ 0\ 1\ 0} +$$

1 1 1 0 1 0 1	PARTIAL SUM NUMBER ONE
1	CARRIES FROM FIRST HALF-ADDITION
1 1 1 0 0 0 1	PARTIAL SUM NUMBER TWO
1	CARRIES FROM SECOND HALF-ADDITION
1 1 1 1 0 0 1	SUM

The sum is the binary equivalent of (denary) 121.

If you look at the first three lines of the sum, you can detect a simple set of rules at work for binary addition:

0 + 0 gives 0

0 + 1 gives 1

1 + 1 gives 0 plus a 1 in the next column as a carry

This kind of rule can easily be applied with logic gates, and the function can be implemented with an AND gate with an EX-OR gate (see Figure 23.1).

fig 23.1   *a half-adder*

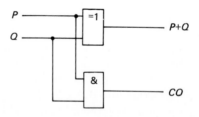

In the diagram, P and Q are the inputs, $P+Q$ is the sum, and CO is the carry to the next column. Two half-adders can be combined to produce a *full adder*, which carries out the two operations, adding the two numbers and also the carry from the half addition. The circuit will, of course, accept a carry from a preceding stage, if there is one. To implement this, two half-adders and another OR gate are required (see Figure 23.2a).

In the diagram P and Q are the inputs, CI_{IN} is the carry from the preceding stage, $P+Q$ is the output, and CO is the carry to the next stage. The full adder is given a simplified symbol, shown in Figure 23.2b. In a real

fig 23.2

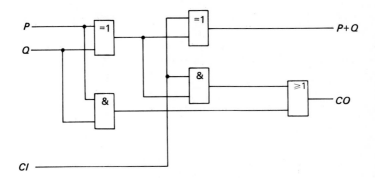

(a) *the gates that go to make up a full adder*

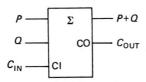

(b) *general symbol for a full adder*

system, such as a computer, 8-bit or 16-bit numbers will have to be added together, so full-adder stages are connected as shown in Figure 23.3; this illustrates a 4-bit adder, for simplicity.

The two 4-bit binary numbers to be added together are gated (i.e. applied via logic gates) on to the adder inputs P_0 -P_3 and Q_0 -Q_3. The sum of the two numbers appears at the outputs $P+Q_0$-$P+Q_3$, and if there is a carry out of the 4-bit sum, the carry appears as a logic 1 on C_{OUT}. Note that C_{IN} is connected to 0 (no carry) if there is no preceding stage. This type of adder is called a *parallel adder,* because all the bits in the number are added together simultaneously. It is a fast method of adding binary numbers, and is the method used in computers. There are certain modifications that can be made to increase the speed even further, but such systems are well beyond the scope of this book.

The other sort of adder is a *serial adder.* The same full adder is used, but the two binary numbers are pushed through, one digit at a time, out of shift registers (rather like a sausage-machine). Each pair of digits is added, and any carry is delayed until the next pair of digits is processed. The shift registers are all controlled by the same *clock pulse* (as we saw in Chapter 21), and a single *D*-type flip-flop (see Figure 21.12, p. 304) operated from the clock pulses will operate as a suitable delay.

fig **23.3** *a 4-bit parallel full adder*

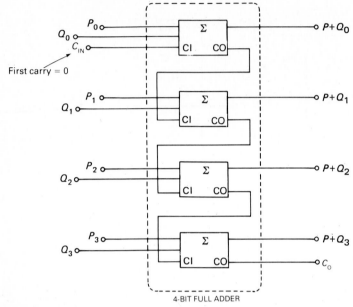

Serial adders are much slower than parallel adders, but can process binary numbers of any length. A serial adder is shown in Figure 23.4.

At last the advantages of binary arithmetic are clear—electronic addition of even large numbers is fast and easy to implement. Subtraction is a little more complicated, but still very easy.

fig **23.4** *a serial adder*

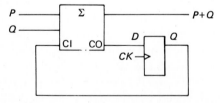

Binary subtraction is accomplished by a process of first converting the number to be subtracted into its negative form, and then adding them together. To convert a denary number into a negative value, we have to subtract it from zero. Thus the negative form of 42 is:

$$0 - 42 = -42$$

Converting a binary number into a negative form is done by producing what is called the *2's complement* (two's complement). This is done in two stages:

1. All the bits in the number are reversed, 0 for 1 and 1 for 0.
2. 1 is added to the lowest bit.

As an example, we can subtract 18 from 103 (they were added above, so it is possible to compare the calculations). The binary equivalent of 18 is 10010, *but*, when taking the 2's complement, we must consider *all* the values stored in the computer's register. Assuming that it is an 8-bit register, the contents will actually be 00010010, the first three 'leading zeros' being just as important as the rest of the number. In the same way, the binary equivalent of 103 is stored as 01100111, with a leading zero to complete the eight bits (which must, of course, be either zero or one).

Using the rules above, the 2's complement of the equivalent of 18 is:

```
0 0 0 1 0 0 1 0    BINARY EQUIVALENT OF 18
1 1 1 0 1 1 0 1    REVERSE BITS
              1    ADD 1
─────────────────
1 1 1 0 1 1 0 0    PARTIAL SUM NUMBER ONE
              1    CARRY FROM FIRST HALF-ADDITION
─────────────────
1 1 1 0 1 1 1 0    PARTIAL SUM NUMBER TWO
                   (NO) CARRY FROM SECOND HALF-ADDITION
─────────────────
1 1 1 0 1 1 1 0    2'S COMPLEMENT
```

Now we can add the 2's complement of 18 to 103:

```
  0 1 1 0 0 1 1 1    BINARY EQUIVALENT OF 103
  1 1 1 0 1 1 1 0    2's COMPLEMENT OF BINARY EQUIVALENT OF 18
  ─────────────────
  1 0 0 0 1 0 0 1    PARTIAL SUM NUMBER ONE
  1 1     1 1        CARRIES FROM FIRST HALF-ADDITION
  ─────────────────
  0 1 0 0 0 1 0 1    PARTIAL SUM NUMBER TWO
1       1            CARRIES FROM SECOND HALF-ADDITION
─────────────────
1 0 1 0 1 0 1 0 1    SUM
```

The final carry has nowhere to go in the 8-bit register, and so is lost.

Binary 01010101 is (denary) 85, which is the correct answer to the original subtraction calculation, 103 − 18 = 85.

In implementing the system in digital logic, NOT gates (inverters) are used to reverse the bits. Adding the 1 to the lowest bit is carried out equally simply, by feeding a carry into the first carry input—held, you will recall, at logic 0 (applied to C_{IN}) during addition.

A 4-bit binary parallel subtracter is shown in Figure 23.5.

It is straightforward to design a logic system that will add or subtract. Serial subtracters are available, but are of course slower than parallel subtracters.

fig 23.5 *a 4-bit parallel subtracter*

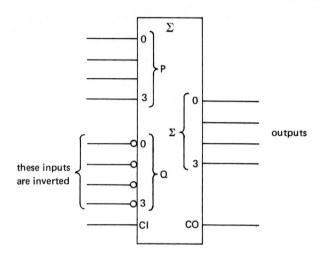

It is possible to design systems for binary multiplication and division, and such systems are sometimes used in digital devices. However, the implementation is rather complicated and, at least in computers, the process of repeated addition is used instead of multiplication, and the process of repeated subtraction for division.

This part of *Mastering Electronics* is not intended to be a comprehensive description of arithmetic circuits, but rather to illustrate the way that binary numbers can be manipulated arithmetically by relatively simple logic systems. The operation of calculators and computers depends on this type of circuit. For a fuller description of logic circuits of all types, including arithmetic circuits, see *Digital Electronic Circuits and Systems* by N. M. Morris (Macmillan, 1974).

23.2 ALGEBRA FOR LOGIC

Before leaving the topic of logic gates, mention should be made of techniques that are available for the minimisation of networks. Systems consisting of a relatively few logic gates can be designed without too much trouble, and where the system can be simplified it is usually obvious. Where large numbers of gates are involved, it is often far from obvious, and formal methods have been developed to help.

Often writing down the *truth table* for the system is enough, but an algebraic method, using a system known as *Boolean algebra*, provides a better method.

An alternative procedure in which all the conditions possible in the truth table are mapped visually can lead to a more rapid solution. The technique is called *Karnaugh mapping*.

Both systems are discussed in detail in more specialised books than this one, including *Digital Electronic Circuits and Systems* by Noel Morris.

23.3 COMPUTER MEMORY CIRCUITS

One of the requirements of a computer system is circuits that will store very large amounts of binary data. The computer memory has to hold many 8-bit or 16-bit binary numbers, available for instant recall by the computer. Even a small personal computer could have a memory system capable of holding at least 390 000 binary digits (bits), so clearly the circuits used have to be economic.

Early computers used a system called *core storage* in which each bit was stored in a ferrite ring, which could be magnetised (1) or not magnetised (0). Core storage was bulky and expensive, each ferrite ring having to be threaded on to a system of wires by hand, rather like weaving. Inside the computer's arithmetic unit, numbers were held in bistables.

With the coming of microelectronics it became possible to use bistables for *all* the computer's memory, and ferrite cores are now found only in museums—just a couple of decades after they were introduced.

The main memory of a computer is called the *random access memory* (RAM). An array of bistables is organised so that any one bistable can be separately accessed, or *addressed*. A standard small RAM chip would hold 1024 bits, organised in a two-dimensional array 32 x 32. This is shown in Figure 23.6.

fig 23.6 *a two-dimensional array that can access 1024 bits*

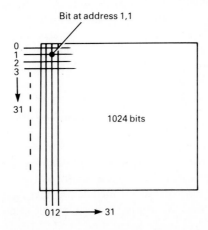

A 10-bit binary number is sufficient to define any one of the 1024 bits. There are 32 possible combinations of 0s and 1s in a 5-bit binary number (2^5) and two 5-bit binary to 32-line decoder systems can be used to obtain the required two-dimensional co-ordinates (see Figure 23.7).

fig 23.7   *two 5-to-32 line decoders are used to address the 32 × 32 array by means of binary numbers*

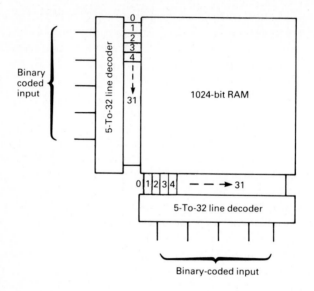

Additional logic is used to (i) set or unset (1 or 0) the bistable, according to the state of an input line (one for the whole array); or (ii) read the state of the bistable and apply it to the data line. The 'set/unset' or 'read the state of the bistable' selection is controlled by the state of an input to the system called the READ/WRITE input. The data input/output is called the DATA line. There is also a third control line called CHIP SELECT, which can disable the entire read/write system. A block diagram of the whole system is shown in Figure 23.8.

The system is intended to provide a memory of 1024 bits for a single position in an 8-bit (or 16-bit) binary register. There is only one data input/output, but according to the address that appears on A_0 to A_7 (Figure 23.8) that data line can be directed to any of the 1024 bistables.

A complete memory for a computer with an 8-bit binary input/output can be produced by using eight identical circuits like the one in Figure 23.8, known as a *1024 × 1 bit RAM*. All the address lines for the eight systems are simply paralleled so that corresponding elements in the eight memories

fig 23.8 *system diagram for a 1024 × 1 bit RAM*

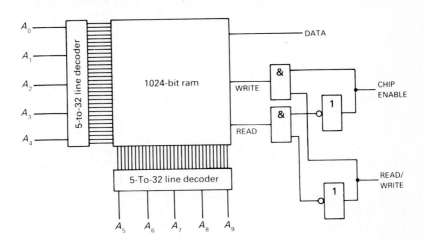

are addressed at the same time. The data input/output lines of the RAMs are connected to different positions in the 8-bit computer data input/ output to provide 1024 8-bit binary numbers. The READ/WRITE lines are paralleled as well, so the computer can either *read* the memory, one 8-bit number at a time, or *write* data into it.

The 1024 × 1 bit RAM is rather small these days. Most computers use chips that have more memory on them. For example, the 6116 CMOS RAM provides 2048 × 8 bits (16 384 bistables) on a single chip. RAM chips containing 8192 × 8 bits (65 536 bistables) or 65 536 × 1 are used in modern designs.

The usual type of RAM is the *static RAM*, available in TTL or CMOS form. Both types are widely used; the CMOS scores, as usual, in having very low power consumption, but the TTL has a faster *access time*, the time it takes to read or write information into the memory. The access times vary, but are in the region of 100–500 ns, with the TTL at the fast end of the scale.

A second type of RAM, *dynamic RAM*, is also used. Instead of using bistable elements, dynamic RAM stores its 1s and 0s in the capacitance that appears as a byproduct of an IGFET's construction, between the gate and source. Unfortunately, the charge in this capacitor gradually leaks away, and the memory cell must be *refreshed* by inputting the data again, every millisecond or so. Special refresh circuits are needed for this, and if the memory is used with a computer, care must be taken to ensure that the refresh cycle is continuous, and does not clash with the computer using the memory.

Despite these disadvantages, dynamic RAM is used in many systems. The advantages are that it is quite a lot cheaper than static RAM, has a short access time (better than 150 ns) and, being simpler, can pack a lot on to a chip.

Addressing and data input/output lines are the same on static and dynamic RAM chips, and the computer cannot 'see' any difference in a properly designed system.

If the memory is to be extended beyond the capacity of the individual chips (i.e. more than 1024×8 using the chip in Figure 23.8), then the CHIP SELECT lines are used. The ten address lines can be added to, and the extra lines used to select groups of memory circuits. Using a 16-bit addressing system, a computer has a further six address lines that can be decoded, giving a further 2^6, or 64, possible combinations that can be decoded. With a 16-bit address, the maximum number of addresses is 2^{16}, or 65 536. A computer with a 16-bit address system could therefore select, in random order and at less than 200 ns notice, any one of 65 536 8-bit or 16-bit binary numbers from its RAM.

For reasons that will become clear in Chapter 24, it is necessary to have a memory system that will not lose its contents when power is removed from the circuit. This is needed to hold a set of permanent instructions for the computer. Clearly, both static and dynamic RAM will lose the contents of the memory when the power is interrupted.

A memory in which data are fixed is called a *read-only memory* (or ROM). The simplest—and cheapest—kind of ROM is the *fusible-link ROM*. This is a system that is addressed in exactly the same way as the RAM in Figure 23.8, but the bistable elements are replaced with diodes, connected across the array as shown in Figure 23.9 overleaf.

Each diode is connected by way of a very thin metal link. As manufactured, every memory location is read as a 1. the ROM can be 'programmed' by applying a pulse of current to the relevant data lines—the current burns through the thin link (like blowing a fuse) and permanently open-circuits that particular cell, which, for ever after, will be read as a 0. Purpose-built instruments are available for 'blowing' ROMs at high speed. Once the required pattern has been impressed on the ROM, it is fitted in the required location in the computer memory addressing system. The read and write currents available from the computer system are insufficient to affect the ROM. Data can be read out of it, but attempts to write data into it have no effect.

Where very large numbers of components are required, a manufacturer may specify a special mask to produce the ROM with the program 'built in' at the design stage.

A compromise between the immutable ROM and the volatile RAM has been produced, in the form of the *erasable programmable read-only memory*

fig **23.9** *an array inside a fusible-link EPROM*

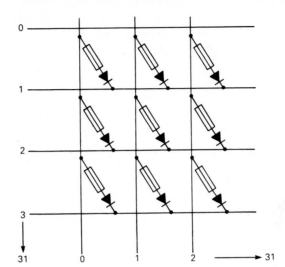

(EPROM). The EPROM uses MOSFET technology, and during program-ming—by the repeated application of pulses of moderate voltage (a few tens of volts)—charges are built up in the insulation below the gate. This layer has an extremely high resistance, and the charge cannot leak away. The presence of the charge causes the MOSFET to conduct, and the charged cell is read as a 1. EPROMs can retain their data for many years. It is, however, possible to discharge the cells in the EPROM by exposing it to intense ultraviolet light. The light causes a photoelectric current to flow, conducting the charge away. EPROMs of this type have a transparent window in the encapsulation above the chip, to allow the chip to be illumi-nated while protecting it from mechanical damage.

QUESTIONS

1. Add the following two binary numbers:

$$100011 \qquad 1001$$

2. What is meant by (i) RAM, (ii) ROM, (iii) EPROM?

3. Convert the following binary number into 2's complement form:

$$10010101$$

4. What is the maximum number of addresses possible using twelve address lines?

5. Dynamic RAM is often preferred in personal computer systems; compare dynamic RAM with static RAM. How do the differences affect the user of the computer?

MICROPROCESSORS AND MICROCOMPUTERS

The first electronic computer to use binary coding was EDVAC, built after the Second World War. It operated on binary numbers up to forty-three digits long, and was able to store, electronically, over 1000 such numbers. It could add, multiply, subtract and divide at the then astounding rate of hundreds of calculations per second. It also used as much power as a small street of houses because it contained literally hundreds of valves. It went wrong once every few minutes of operating time, mainly because of the inherent unreliability of valves and because of the high voltages involved. Needless to say, it was hugely expensive to build and run. It was however the grandfather of modern compact, efficient computers and used the same kind of organisation, even down to magnetic tape as a bulk storage medium.

In order to reduce to a minimum the number of interconnecting wires required, all computers use a system in which *buses* carry information from one place to another inside, and sometimes outside, the computer. A bus is nothing more complicated than a group of wires, used together for the transmission of binary numbers. In Chapter 23 you saw the way in which binary numbers can be represented electronically – so a bus consisting of 8 wires (known as an 8-bit bus) will carry an 8-bit binary number – any number from 0 to 255. A 16-bit bus will carry a number between 0 and 65 535, and a 32-bit bus will carry any number from 0 to 4 294 967 295! Small personal computers tend to have an 8-bit bus for data, and a 16-bit bus for addresses. 'Data' and 'addresses' need explaining. Figure 24.1 shows a system diagram for a typical computer.

The first main component in the system is the *central processing unit* (CPU). The CPU performs all the calculating, organising and control functions, and is often called (inaccurately) the 'brain' of the computer. The CPU is connected to the rest of the system by three groups of wires: the *data bus*, the *address bus* and the *control bus*. Also connected to the buses are the *memory* and an *input and output* system.

fig **24.1** *system diagram for a microcomputer*

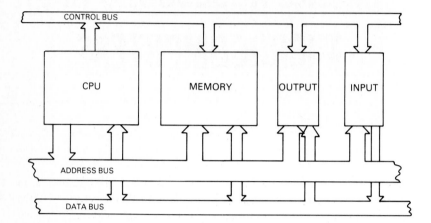

The data bus is used by the system to transmit binary numbers (data) from point to point. Data may be part of a calculation, or an instruction (more about that below). An 8-bit bus does not limit the computer to dealing with numbers less than 255, of course. Larger numbers are simply sent in two or more pieces, one after the other. Numbers up to 65 535 can thus be sent in two pieces, called the low-byte and the high-byte. A 'byte' is simply eight bits (it is a contraction of 'by eight') and is, in small computers, the amount of information that is dealt with at one time by the CPU.

The address bus is used by the CPU (or other devices) to look at any indivudal byte of memory (either **RAM** or **ROM**); in small computers, a 16-bit address bus is used to provide a total memory capability of 65 535 bytes. This is referred to (confusingly) as '64 kilobytes', or '64K', the capital **K** being used to denote the 'computer kilo' of 1024.

The memory itself consists of a large number of bistable memory elements – described in Chapter 23 – along with appropriate decoding circuits. The memory is used for storing data (information processed by the computer) and program (instructions telling the CPU what to do). Input may consist of a keyboard, or anything else that can feed information into the computer. Output may be a printer, visual display unit (VDU), or some other device enabling the computer's output to be passed on to the outside world.

24.1 CPU OPERATION

When the CPU is first switched on, it sends a signal down the address bus, corresponding to a 'one' in binary, 00000001. It also sends a signal down one of the two most important lines of the control bus, the READ line. At the same time, it prepares to receive a number down the data bus, and 'looks at' the data bus for information. The memory-decoding circuits connect the relevant memory bistables to the data bus, so that the contents of the bistables appear on the data bus. The CPU then loads the contents of the data bus into its own circuits, and the first byte of data is read into the CPU.

It is crucial to an understanding of computers to know that what happens next depends entirely on the contents of the byte of data.

Although in a large computer the CPU may be made with large numbers of individual ICs, the CPU of a small computer will almost certainly be a single chip. This chip is called a *microprocessor*. Any computer which has a microprocessor as its CPU is called a *microcomputer* (regardless of its overall physical size). The microprocessor is a tremendously complicated piece of circuitry (see Figure 11.8) and can recognise the number that appears on the data bus. Since it is better to use a real-life example, we will use the Z80 CPU, manufactured by Zilog – it is probably the most popular of all microprocessors to date, having been widely used in many small computers including the best-selling Sinclair Spectrum. The Z80 can recognise almost 200 basic types of instruction. Assume that the byte just loaded from memory location 00000001 was 62. This is interpreted by the Z80 as an instruction. It tells the microprocessor that whatever is in the next memory location should be loaded into an 8-bit storage area (called a register) called *register A*. This done, the Z80 puts a new address on the address bus, 00000010 (2 in denary), looks at the data in that memory location, and loads it into register A.

The Z80 now moves to memory location 00000011 (3 in denary: to prevent this book becoming too long, we will use denary instead of binary from now on!). It expects another instruction: it might be to load register B with the next number, represented in Z80 code as 6. On to memory address 4 to find the number itself, then to location 5 for the next instruction. It might be 128, which means 'add the contents of register A to the contents of register B and store the result in register A'. Already, we have written a very simple computer program, to add two numbers:

    Load A with a number
    Load B with a number
    Add A to B and store the result in A

This can be written more succinctly in a programming language called 'Z80 Assembly Language'. It looks like this:

```
LD A,2
LD B,3
ADD A,B
```

This, as you can see, adds 2 to 3, leaving the answer, five, in register **A**. In this example, the numbers to be added are immediately following the instructions, but they need not be. For example:

```
LD A, (30000)
LD B, (30001)
ADD A,B
```

takes the numbers from memory locations 30000 and 30001, regardless of whereabouts in memory the program has been put. Remember that the CPU can treat the contents of any memory address as data or as program, according to the context.

Assembly Language is not itself recognised by the microprocessor, since Assembly Language uses letters and numbers; the microprocessor recognises only binary data between 0 and 11111111. But it is convenient to write programs in Assembly Language as that is easier for humans to follow. . . try to imagine a program consisting of, say, ten thousand binary numbers, then try to imagine finding a mistake in it! Assembly Language can be translated automatically into binary numbers (called the object code) by a suitable computer program, called an Assembler.

The CPU can do much more than merely add numbers. It is worth looking in a little more detail at the way a typical microprocessor is laid out, and how it works. The Z80 makes a good example.

The Z80 has 22 registers, including register **A** and register **B** that have already been mentioned. Remember that each register is actually a series of 8 bistables, and is capable of holding one byte. The purpose of many of the registers is too obscure to go into in a general book on electronics, but the more obvious ones are the main data registers of which there are seven of them, called **A, B, C, D, E. H,** and **L**. There is also a second set of registers, identical to the first, called the alternate register set. They are called **A′, B′, C′, D′, E′, H′,** and **L′**. There is also the PC, or program counter register – this is a 16-bit register. As we have seen, the registers are used for temporary storage of data. We can best see how they are used by means of an example program. For simplicity, we will ignore the input and output side of things, and assume that the computer starts off with a number in the **A** register, and that placing the answer in the **A** register at the end of the program is what is required. The following program is intended to carry out the following task:

1. check to see if the number is 10. If it is, simply put the 10 in the **A** register and stop.
2. If the number is 20, then the computer is to begin a program that starts at memory location 35000.
3. Any other number is to be doubled, and the answer put in the **A** register.

Here it is:

```
CP 10
RET Z
CP 20
JP Z,35000
LD B,A
ADD A,B
RET
```

Operation of the program is simple enough. CP 10 Com**P**ares the **A** register with the number, in this case 10. It does this by experimentally subtracting 10 from the **A** register. The next instruction is **RET**urn if **Z**ero; so if the result of the experimental subtraction is zero, the **RET**urn instruction terminates this part of the program by returning to the point at which the program was started (or 'called'). This accomplishes the first part of the program specification.

The third and fourth lines of the program work in a similar manner, except that the **J**um**P** if **Z**ero to 35000 instruction sets the PC register – the program counter – to 35000. This has the effect of interrupting the orderly sequence in which the Z80 looks at each memory location in turn, forcing it to carry on from location 35000.

The last part of the job is accomplished in the next three lines. **L**oa**D** register **B** with the contents of register **A** is followed by **ADD** the contents of register **B** to register **A**. This is a way of doubling the size of the number in **A**. The final instruction is **RET**urn, which ends this part of the program.

This is a very simple example. Real programs would be much longer, and might use all the registers. You get some idea of the speed at which the Z80 works when you know that the program above takes a maximum of 20 microseconds to run.

The most important things to understand at this stage are:

1. the CPU is able to look at any of the memory locations, but in the absence of any instructions to the contrary will look in the memory locations sequentially immediately it is turned on;
2. the contents of each memory location (8 bits, or one byte) are read by the CPU, and what the CPU does next depends on the contents of the byte it has just read;
3. memory locations can be used for storing two sorts of information: the program that controls the activities of the CPU, and data. Data may be data that has been supplied to the computer from some external source,

or it may have been put there by the CPU. There is no difference at all between the memory locations used for these two different purposes.

24.2 SYSTEM OPERATION

In looking at the computer, it is convenient to look at the individual components of the machine – the CPU, memory, keyboard, etc. But in order to make any sense of the computer, it is important to understand the way the whole thing operates as a system.

Figure 24.1 is a very important diagram, as it shows the way the system works. The three buses are the channels through which the parts of the computer communicate with each other, and every computer is organised in much the same way. Almost all computers are designed in such a way that extra components can be added to the system, simply by plugging them into a multi-way socket that connects directly to the three buses.

The minimum components are the CPU and a memory. The lower addresses of the memory are generally ROM, so that immediately power is applied to the CPU it starts executing a program, bringing the system to life. The CPU can be directed by the program to look at an input device. The most common input device is the keyboard. The keyboard may itself incorporate circuits that translate the key pushes into the relevant binary code. So pushing a letter 'e' would generate a binary number 10100110, to be applied to the data bus at the right instant and read into the CPU. Alternatively the CPU itself could run a short program, itself part of a larger program, designed to look at the connections to the keys, thus saving on hardware. Such a program is called a subroutine, and saves electronic components at the price of a small loss of operating speed.

Output devices also work by using the buses. The circuits driving a VDU can interrupt the CPU briefly to examine the relevant memory locations for the screen, then interpret them as characters or graphics shapes for display on the screen; the way in which the 'screen memory' is addressed is exactly the same method as the CPU uses, but the address and data buses are temporarily under the control of the VDU system.

At this point it is interesting to consider the way the memory locations are translated into a picture on the screen, and to examine what is known as the *memory map* of a typical computer. The memory map simply shows the allocation of the addresses that the CPU can access. This is the memory map of a typical low-cost microcomputer than can address 64K (65536 bytes).

The first ('lowest') 24K is occupied by the ROM. The ROM contains the *operating system*, the programs that do the housekeeping – all the operations of looking at the keyboard, loading and saving programs, controlling input and output, etc. The ROM also contains a BASIC *interpreter*

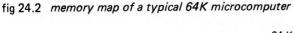

fig 24.2 *memory map of a typical 64K microcomputer*

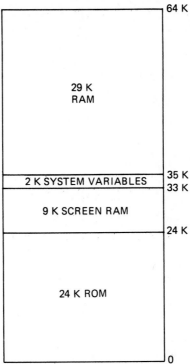

that enables the computer to run programs written in that language – more about that later. Remember that the microprocessor always starts at address 0 when switched on, so the ROM is always at the bottom of memory. The entire memory map above the ROM is filled with RAM – 40K of it. Parts of the RAM are reserved for special purposes. The 9K between 24K and 33K is the screen RAM and is used to store the information which the computer translates into a picture on the screen of the monitor or TV. Figure 24.3 shows the way the information is held.

The screen is divided into 24 lines of 48 characters, each character requiring eight bytes. The total RAM required for this screen is therefore 24 × 48 × 8 = 9216 bytes, or 9K. The bits in each group of bytes representing a character store the shape of the character itself, illustrated in Figure 24.4.

As an illustration of how the computer works, consider what happens when you type a letter (make it an 'E') on the keyboard. To begin with, the microprocessor is running a program that is part of the ROM, checking the keyboard at regular intervals to see if a key has been pushed. At the

fig 24.3 *the way information can be held in the computer's RAM, for display on the screen; this part of memory is sometimes called the 'display file'*

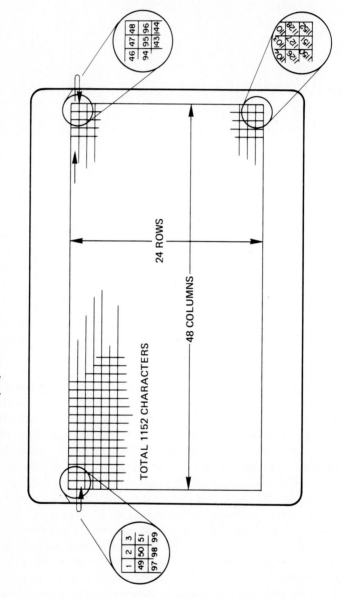

24 ROWS

48 COLUMNS

TOTAL 1152 CHARACTERS

same time the VDU system is reading the screen **RAM** and displaying it on the screen as a picture. You press the key. The microprocessor checks which key has been pushed and, according to which **ROM** program it is running, may store the result in **RAM** somewhere. The microprocessor next uses a look-up table, stored in the **ROM**, to find out which codes are required to produce the letter 'E'. It then checks other **RAM** locations to

fig 24.4   *the bits of eight bytes are used to represent a character on the screen, like this 'E'*

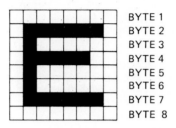

BYTE 1
BYTE 2
BYTE 3
BYTE 4
BYTE 5
BYTE 6
BYTE 7
BYTE 8

| 0 | 0 | 0 | 0 | 0 | 0 | 0 | 0 | 0 | I | I | I | I | I | I | 0 | 0 | I | 0 | 0 | 0 | 0 | 0 | 0 | 0 | I | I | I | I | I | I | 0 | 0 | 0 | I | 0 | 0 | 0 | 0 | 0 | 0 | etc |

   BYTE 1   BYTE 2   BYTE 3   BYTE 4   BYTE 5

see whereabouts on the screen the letter should be printed, and finally writes the eight bytes that represent the letter into the relevant places in the screen **RAM** – note that the locations are not consecutive, but are actually 48 bytes apart. See Figure 24.5 on page 362. This will result in the letter 'E' appearing.

In practice the computer has to do a lot more than this, since the 'E' may or may not have significance in the context of the program that is being run.

Above the screen **RAM** (usually) there is another area of workspace that the microprocessor uses for storing temporary data like the printing position on the screen, the intermediate results of calculations, and all sorts of pieces of information that the system needs to remember. In this memory map, 2K of RAM is allocated. This leaves 29K of 'user RAM' available for programs. Which brings us to programming languages.

24.3 COMPUTER LANGUAGES

The CPU of a computer – whether in a microcomputer or the largest mainframe – is programmed in binary code. It is pretty well impossible for humans to use binary code for programming. The nearest usable language

fig 24.5 *the way characters are displayed on the screen of the monitor*

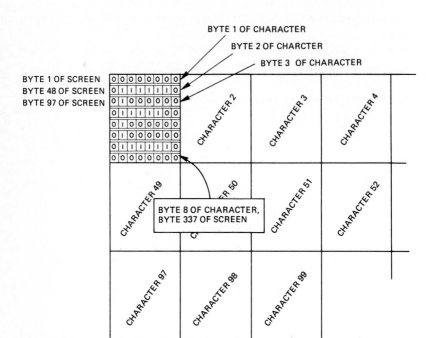

to the binary code the CPU needs is Assembly Language. Assembly Language instructions have a one-for-one correspondence to machine instructions: in other words, each Assembly Language instruction has an exact equivalent in binary code. Assembly Language is not easy to learn, and it takes a long time to program a computer to do anything useful. An Assembly Language program to input two six-digit decimal numbers and divide one into the other, expressing the result as a decimal number, would take an experienced Assembly Language programmer a full week to write. Clearly there needs to be an easier way.

Assembly Language is known as a *low-level language* because it is close to machine language. There are other computer languages that are much nearer to English, and are consequently easier to learn. Such languages can make it much simpler to program a computer, and are used wherever possible. Such computer languages are called *high-level languages*.

There are two classes of high-level language, *compiled languages* and *interpreted languages*. Both translate something closer to English into a code understood by the CPU, but they do it in different ways. We will start by looking at the most widely used computer language of all, BASIC.

The name is an acronym for **B**eginners **A**ll-purpose **S**ymbolic **I**nstruction Code, and it was first used in the USA for teaching programming to university students, but has since been developed and extended until it can be used for a wide range of programming applications. BASIC is an interpreted language. A long and complex program (written in Assembly Language!) is kept in the ROM – this program is the BASIC Interpreter, and translates a program written in BASIC language into the binary code that the CPU requires. Here is a simple BASIC statement:

```
10 LET A=25.4/16.32
```

It is fairly obvious that this program line instructs the computer to let **A** (a variable, as in algebra) equal 25.4 divided by 16.32. It takes a few seconds to write such a line – but, via the interpreter, it makes use of machine code that may have taken months to write and refine. The line is typed into the computer, where it stored in RAM. The computer does not actually use the program line until instructed to do so. The instruction to start a BASIC program is RUN. When the computer receives this instruction it looks at the program line, and, a piece at a time, refers it to the interpreter. Each instruction is translated – as the program runs – into machine code that the CPU can understand. The interpreter does this by making use of an array of machine code programs in the ROM, calling each program into use as required.

The program line has a number (in this case, 10). Other program lines can be added, with higher or lower numbers, and the computer will sort them into order and run them sequentially. Here is a three line program:

```
10 INPUT a
20 LET b=a/4
30 PRINT a,b
```

When this program is RUN, the screen responds with some sort of 'prompt', perhaps a question mark. This means that the computer is waiting for an input. Type in a number, for example, 42. Almost immediately, the computer prints on the screen:

```
42        10.5
```

You should be able to see quite easily how the program works. . . BASIC is as simple as that, at least for uncomplicated programs.

Here are two more BASIC programs, the second more complicated than the first. It is one of the most important features of BASIC that it is fairly obvious what they do, even if you are relatively new to programming:

```
10 PRINT "Type in the radius"
20 INPUT r
30 LET a=PI*r^2
40 PRINT "The area of the circle is ";a
50 STOP
```

```
10 REM Multiplication Tables
20 INPUT a
30 FOR i=1 TO 12
40 PRINT a;" times ";i;" is ";a*i
50 NEXT i
60 STOP
```

Another advantage of BASIC is its ability to detect problems in your programs. The interpreter, in translating the program, carries out many checks on whether or not the program makes sense. If it does not, the program stops running, and the interpreter generates an error message. Here is a BASIC program with a mistake in it:

```
10 INPUT a
20 LET b=a/0
30 PRINT b
```

When this program is RUN, the computer stops with the message:

```
DIVIDE BY ZERO ERROR AT LINE 20
```

The programmer has inadvertently asked the computer to divide a number by zero, which cannot be done, and the interpreter has detected it. Such an attempt, if made when using a program written in Assembly Language, would not be trapped, and the results would be unpredictable – more important, a machine-code routine with this sort of error might well overwrite itself, and would have to be reloaded.

The great snag with BASIC – and all other interpreted languages – is that by computer standards it is *slow*. The interpreter will spend most of its time translating the BASIC and choosing the relevant machine-code routines. It also spends a lot of time checking for errors. The result is that BASIC runs between 50 and 200 times slower than well-written Assembly Language, depending on what it is doing. For some purposes – such as stimulations, business and scientific applications requiring the manipulation of very large amounts of data, and arcade games, BASIC is not fast enough. The next step upwards in speed is to use a compiled language.

BASIC is usually an interpreted language, that is, a language in which each instruction in the program is translated into machine code (by calling on a bank of stored machine-code routines) as the program is running.

Since the translation takes up most of the time, it follows that a big increase in the running speed of any given program could be obtained simply by carrying out all the translation first. This is the essence of a *compiled* language. The program is typed into the computer and stored. When the program is finished, the computer is given an instruction to compile it. The *compiler* (equivalent to BASIC's interpreter) translates the program into machine code, which is also stored. The typed-in program is called the *source code*, and the compiled version is called the *object code*. Notice that the program is 'translated' completely, before it is run. It is during this translation phase that any error messages are generated.

The source code is now no longer needed – it can be stored away on disk or tape (see below) for future reference. The object code can now be run by the computer, and since the translation has already been carried out, execution of the program is much faster.

There is no reason why **BASIC** cannot be compiled. Indeed, many **BASIC** compilers have been written and are commercially available. There are however a number of features of **BASIC** that make it a less than ideal language for treatment in this way; it was, after all, intended to be used as an interpreted language. Of all the compiled languages, Pascal is one of the most popular. The name is not an acronym this time, but is a tribute to Blaise Pascal, a seventeenth-century mathematician and philosopher. Pascal was designed at the outset to be a compiled language, and also to have a form such that its users are almost forced to write programs in an orderly, understandable way. Pascal compilers do not actually compile directly to machine code. Instead, they compile into an intermediate form called a *P-code*. The P-code is itself then run as an interpreted 'language', using a P-code interpreter! But the 'interpreter' is generally called a *translator* in this context, and the result is something that runs a lot faster than an interpreted language, because all the hard part of the translation (Pascal to P-code) is done before running the program.

The speed of a compiled language is a function of the quality of the compiler – all else being equal, the better the compiler, the faster the object code will run. The skill in writing a compiler is in getting it to produce a relatively economic code. Although it is beyond the scope of this book to deal with programming languages in any detail, here is a short program segment designed to accept a 'y' or an 'n' (for yes and no) and nothing else as an imput, first in **BASIC**:

```
10 INPUT a$
20 IF a$="y" OR a$="n" THEN  GO TO 50
30 PRINT "Try again!"
40 GO TO 10
50 PRINT "OK"
```

. . .and in Pascal:

```
REPEAT
  READ (answer);
  IF (answer<>'y') AND (answer<>'n') THEN WRITE ('Try again!');
UNTIL (answer='y') OR (answer='n')
WRITE ('OK')
```

There are, of course, many different high-level programming languages. They are all easier to write than Assembly Language, and they all run slower, for no compiler or interpreter has yet been written that can equal well-written Assembly Language for efficiency. Programming computers is something people can still do better than computers!

One of the oldest programming languages is FORTRAN (**FOR**mula **TRAN**slator). It is still an excellent language for science and mathematics, and bears a close similarity to BASIC, which was developed from it. Another language that is still widely used is COBOL (**CO**mmon **B**usiness **O**riented **L**anguage) which is good for producing lots of long reports, inventory and stock control, but too 'wordy' for scientific work, graphic programs, or mathematics. Pascal itself is a good general-purpose language, but is not particularly good for control applications. For heavyweight applications – defence networks, for example – languages like FORTH and Ada are used. For experiments in artificial intelligence (trying to make a computer behave like a person) a language called LISP is often used. For applications programming where transportability (jargon for ease of translation for different makes of computer) is important, languages called BCPL and C are currently popular. All in all, there are enough computer languages to fill a tower of Babel – all with advantages and disadvantages that may recommend them in different applications.

To complicate matters even further, many software writers have found it efficient to use hybrid programs that are mostly written in a high-level language like BASIC but have machine-code subroutines (program sections within the main program) for the parts that need to work fast.

24.4 BACKING STORAGE

A computer can store information in **RAM**, and the operating system and interpreter can be written into **ROM**. There is, in addition to this, a need for some means of relatively permanent storage of data or programs – something that will store the information while the computer is switched off.

The simplest way to store such information is on magnetic tape, and even the cheapest home microcomputer has the means to record a program onto tape, and then to retrieve it. Cassette tapes are widely used because of their convenience. It works like this.

Assume, to start with, that you have typed a program into the computer, got it running satisfactorily, and now want to save it. To record the program on tape, you would simply type SAVE "myprogram" on the keyboard, press the 'record' button on the tape recorder, and press ENTER (or RETURN) on the computer. The computer would then transmit the program in a suitable form for recording. When it has finished, you stop the tape – some machines have a built-in relay to do this automatically. A useful feature on some computers is a VERIFY command. You play the tape back into the computer after typing this, and it compares what is on the tape with what is in the relevant part of the RAM to make sure the recording is in order.

There are many different systems used for recording the information, but one of the best (and most commonly used) is pulse width modulation. Very little in the way of hardware is required for this, so it is cheap, too.

When the operating system in the ROM receives the instruction to SAVE a program, it makes several checks. First, it checks to see that there is in fact a program in the computer. No point in saving something that isn't there! Second, it checks to see how long the program is. If the operating system is one in which the program may occupy different parts of the RAM, it also checks to see where the program starts. Finally (in most systems) it works out a checksum for the program – a meaningless number (like the number of binary 1s in the program) which is used to ensure that the program is intact when it is reloaded after saving. These various bits of information are put together into a *header*, a short block of data that is recorded before the program itself.

The header (program name/position in RAM/length/checksum) is recorded first, followed by the bytes that represent the program. The principles of pulse-width modulation were considered briefly in Chapter 22, but for computer programs it is necessary only to differentiate between 0 and 1. Figure 24.6 shows the way the signal is encoded, a long pulse representing a 1 and a short pulse representing 0. The train of square-wave pulses is produced at a frequency within the audio range, so the resulting waveform can be treated just like sound and fed into the recorder's microphone input. The maximum rate at which data can be recorded depends upon the frequency response of the recorder, and in practice about 1500 bits per second is a sensible maximum for reliability. Before any data is saved (even the header) the computer transmits a couple of seconds of a steady tone, to allow the recorder's automatic gain control circuits to settle at the right level. A medium-sized program of 20K would therefore take roughly 112 seconds to record. Manufacturers often use slower rates, out of pessimism about the quality of recorders that will be pressed into use with their computers.

fig 24.6 *binary data encoded for recording on tape*

Once the program is recorded on tape it can of course be stored indefinitely.

The program is recovered by typing LOAD "myprogram", and playing the tape into the relevant socket on the computer. The first part of the program to be received is the header. Once again, the computer makes a few checks. Do the program name typed in after LOAD and the one in the header correspond? If not, the program will not be loaded. Is there room in the computer for the program? Most computers are available with different amounts of RAM, and it is clearly pointless trying to load 40K of program into 32K of RAM. The computer sets the start address if necessary, and loads the checksum into a counter. Now the program is loaded, each 8 bits going into successive bytes of RAM. As the program is loaded, the checksum is decreased by 1 for each bit that is 1. When the program is completely loaded, the checksum ought to have reached zero. If it has not, or if it is less than zero, then something has gone wrong with the recording or playback, and the computer displays a TAPE LOADING ERROR message on the screen.

While recording programs or data on tape is cheap and straightforward, it does have disadvantages. It is not particularly fast – a couple of minutes for a 20K program does not seem too bad, but ten minutes for 100K is long enough to be a nuisance. The limited speed of recording on cassette tape is mostly caused by the fact that small recorders are designed for audio frequencies. If a recorder is designed specially for digital use, the record and playback speed (computer people call it the read and write speed) can easily be pushed up to 50 000 bits per second; at this rate, 100K of data can be transferred in about sixteen seconds.

The second disadvantages is that the recorder still requires some kind of manual intervention, and it would be much better if the computer could control the mechanical part of the recorder: this too is easily fixed by using a special design.

A number of low-cost microcomputers use just this technique... a specially designed cassette tape recorder, either built into the computer itself, or sold as a specialised add-on.

Indeed, the mainframe computers of the 1960s generally used tape as a mass-storage medium for data. The tape drives were very specialised and very expensive. To increase the read/write speed they used wide tape (generally one inch) and at least nine record/play heads, allowing one byte and a check bit to be recorded simultaneously, increasing the speed by a factor of nine. Read/write speeds were also very fast, and computer tape still represents a very compact way of storing data.

Tapes however, do have a major drawback. Suppose the data the computer has to get at is in the middle of a reel somewhere! The tape has to be run forward to the relevant place (which has to be located somehow) before the data can be loaded. If data or program segments are scattered at random along the tape, it is going to take the program a long time to run. To some extent this difficulty can be reduced by careful organisation of tapes, but what is really needed is a storage system that is more like RAM – a system in which the computer can find information almost immediately wherever it is recorded.

24.5 DISK STORAGE

Anyone who has used a tape recorder and a record player will have discovered that it is much easier to find a particular track on a record than it is on a tape. The record is effectively a random access storage system, whereas the tape is not. Computer tapes and disks are an exact parallel.

The cheapest kind of disk is the *floppy disk*. This is a thin circular sheet of flexible plastic, either $3\frac{1}{2}$ or $5\frac{1}{4}$ inches across. There is a hole in the centre, and an outer protective sleeve of some sort. The surface of the disk is covered in magnetic oxide – the same material that is used for the coating on ordinary recording tape. Figure 24.7 shows the layout of a typical floppy disk drive.

The read/write (record/play) head is brought into contact with the surface of the disk when the disk is inserted. The disk is spun round by a small motor, and the head is mounted on an arm that can move it to any point on the disk's radius. The arm is driven backwards and forwards by a stepping motor, a motor that is turned by signals generated by the computer. The computer can position the arm very accurately and can record either forty or eighty tracks across the disk.

Before the disk is used, it has to be formatted. This is an operation done by the computer, checking each of the tracks and dividing them up into sectors by recording marker pulses at various points. One track is reserved for the directory. When the computer receives an instruction to save something on the disk, it first checks the directory to see which parts of the disk are available for recording. The directory contains information about the name of what is stored, how much of it there is, and which parts of which tracks it is on. Having located a clear part of the disk, the computer makes an entry in the directory, then records the information itself.

fig 24.7 *a typical floppy disk drive mechanism, somewhat simplified*

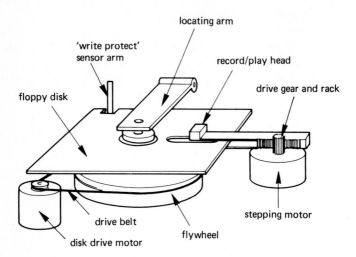

To load the data or program back from the disk, the computer simply looks in the directory to find out where the information is, drives the head to the relevant track, then loads the data from the relevant sectors. The whole operation can be very quick. In practice, the speed of floppy-disk storage varies enormously. A system with an old-fashioned or poorly written disk-operating system might take as long as a minute to find and load a short program, whereas a good system might do it in two seconds!

A floppy disk typically holds between 80K and 800K bytes, according to design (and cost). Improvements are continually being made.

Although the disk drive will contain its own electronics, the computer system operating it must have suitable programs in its operating system. Because of the amount of precision mechanical work – always expensive – a floppy disk drive can cost as much as the microcomputer it serves. But for any serious use, a floppy disk is the minimum storage system required.

It is worth mentioning a system that provides disk-like facilities at a relatively low cost, the stringy-floppy. This oddly-named device uses cartridges containing a short (less than 10 m) length of tape, arranged in an endless loop. Mechanically much simpler than the disk drive, the stringy-floppy has the same kind of operating system as a disk, recording a directory on the tape in just the same way. The difference is that during a read or write cycle the tape whizzes round twice – once to find the directory and look at the entry, and again to fetch the data. If the loop takes 5 seconds to get round, then there is going to be an average 5 second overhead on each operation. Possibly as little as $2\frac{1}{2}$ seconds, possibly as much as 10 seconds. Nevertheless a well-designed stringy-floppy can easily work faster than a floppy disk with a poor operating system. Up to the time of writing, the only really successful stringy-floppy has been Sinclair Research's Microdrive, with an average loop time of about 6 seconds, and a capacity of about 100K bytes. The simplicity of the mechanical components (a greatly simplified cassette tape drive) makes the price less than half that of a comparable floppy-disk.

Really large computer installations, either business microcomputers, minicomputers or mainframes (in ascending order of size) may use *hard disks*. The smallest and cheapest is a design known as the *Winchester disk*. The Winchester disk consists of several solid aluminium disks about half an inch thick and a few inches in diameter. Each disk is a precision component and is made perfectly flat on both sides. In principle it is similar to the floppy disk, but is sealed permanently in a perfectly dust-free chamber. The read/write heads skim across the surface of the disk without quite touching it, so there is no wear. The Winchester disk can typically hold several megabytes (millions of bytes) of data.

The biggests computers use bigger disks still, the 14-inch hard-disk pack. Several disks are fitted together into a 'pack', the whole pack being replaceable. Both sides of the disks are used, and each surface can hold about 30 Mbytes, although this is continually being improved. Hard disk systems are not cheap – currently a hard disk system costs about as much as a 'top of the range' family car.

24.6 MASTERING COMPUTERS

There is space enough in *Mastering Electronics* to give only the briefest outline of the way computers work. Readers wishing to find out more about computing and computer systems should refer to *Mastering Computers* by G. G. L. Wright (Macmillan), and also *Mastering Computer Programming* by P. E. Gosling (Macmillan).

QUESTIONS

1. Draw a system diagram linking the computer's CPU, RAM, ROM, input, and output.

2. Why is binary machine code never used to program a computer, no matter how simple?

3. Most small computers use BASIC for programming. State three important advantages that BASIC has over Assembly Language for programming.

4. Explain the following terms: (i) disk drive, (ii) compiled language, (iii) interpreted language, (iv) 64K RAM, (v) memory map.

5. Why must even the simplest computer be equipped with a ROM?

COMPUTERS, ELECTRONICS AND THE FUTURE

It is clear that the future of electronics, or at least a large part of its future, will be involved with computers. Just as the IC has in many cases replaced complicated circuits using discrete components, so it seems likely that the microprocessor and its offspring will replace whole systems.

The use of a microprocessor to simplify a complex system has already been discussed, but it is equally possible to use a microprocessor to make practicable a system that would otherwise have been too expensive or too large. Digital alarm clocks could have been made in the 1940s, but the time, effort, expense and overall size of the finished product made the project completely impractical. Systems that are today too large or too expensive to be sensible commercially may before too long, become cheap and commonplace as a result of advancing microprocessor technology. It is almost impossible to predict what these systems will be. Who, thirty years ago, would have predicted home computers?

There is also the unexpected technological 'breakthrough'. It is this that really thwarts any attempt at long-term predictions. Such predictions must be based on current technology and likely developments of it. In the 1940s people were predicting 'radio valves the size of a pea' within twenty years. The discovery of the transistor actually resulted in whole computer CPUs substantially smaller than that, and electronics tiny enough to pack a complex multifunction chronometer system into a rather slim wristwatch, along with a battery that powers it for more than five years. The next unexpected discovery may be equally revolutionary.

In the immediate future it seems likely that computers, so far the ultimate in generalised systems, will be very widespread. Small computers are cheap, reliable and flexible enough to be used in applications ranging from interactive games (chess, backgammon, adventure games, 'reaction' games like the Space Invaders derivatives) to automatic production control. The index for this book was produced using a very small 'home computer'.

There is also the generally predicted rise in what has been termed *information technology*. Information technology is the transmission of information of all kinds—pictures, text, numbers—through electronic rather than mechanical means. Large *data bases*, pools of information held in computer systems, generally on disk, will be available to people and organisations, and this should make for greater efficiency in domestic business life. Already the various teletext systems enable people quickly and cheaply to reach information held in a computer data bank. Computers linked to video systems seem to have great promise for the future, though it is commercial pressure that dictates the lines of development for technology. If there isn't a market for it, there is unlikely to be money available for the research.

Satellite television links, in which information from a transmitter on the ground is sent out into space to a communications satellite and then back again, mean that a single broadcast can reach a whole hemisphere of the earth. The social impact of this, particularly in undeveloped countries, may be considerable. The technological impact may be to force the development of inexpensive receiving systems.

Electronics technology is having, and will continue to have, an impact both on work and leisure. At work, microprocessor-based equipment is changing the face of industry. Word processors are more efficient than typewriters for most office work; they also make the job more interesting, and at the same time more taxing. Industrial robots are widely used for assembly work, replacing skilled and unskilled people in various routine tasks. In the short term, this is bad news. Commercial pressures will force companies to install robots, and inevitably people will lose their jobs. In the long term, more is being done for less effort on the factory floor—yet jobs are created in factories where the robots are designed and manufactured. There will be less dull, routine, dangerous and dirty jobs, but more jobs for trained technologists.

This brings us neatly back to the subject of this book. An understanding—or mastery—of electronics is a skill that industry and society will need in the future. The design and production of electronic equipment, or of machinery controlled by electronics, is vitally important for the future, for it is in these areas that the lost production-line jobs will be recovered.

Look at the changes in our society that have taken place as a result of the development of electronics. Look at the effects of colour television, video-recording, hi-fi, digital clocks and watches, calculators and computers, and look especially at the industries that have grown up as a result of these products. Look, too, at the effects of electronics on military technology: it is not now the largest army that wins a battle, but the army that is technologically superior.

Now try to guess how electronics and computers will affect us all in the next thirty years. Read science fiction. It may give you a few clues.

APPENDICES

CONSTRUCTION PROJECT – MODEL RADIO-CONTROL SYSTEM

The radio-control system described in the text makes an interesting and instructive constructional project. It is more complicated than the average 'student construction project', but has the advantages that it pulls together something of radio and digital techniques and the end-result is worth while in itself.

There should be no serious problems if the circuits and instructions given in the text of this book are followed carefully, and with an understanding of the way the systems work.

If reliable results are to be obtained, it is a good idea to follow the circuit layouts given in this appendix quite closely. The printed circuit board patterns are reproduced full size, and many students will want to make their own. However, *all the printed circuit designs given in this appendix are available as ready-made boards, tinned and drilled. They can be obtained from Maplin Electronic Supplies Ltd, P. O. Box 3, Rayleigh, Essex SS6 8LR, England*, who can supply all over the world by mail order.

The boards are designed with rather more 'space' than would be the case for commercial designs, to ease construction. The transmitter, receiver, encoder and decoder are all on separate circuit boards so that they can, if required, be put together by separate project groups.

Also included are oscilloscope photographs, illustrating the waveforms that are to be expected at various points in the circuit. This should help with fault-finding in projects that do not work first time.

This model control system requires no license to operate it in the UK. The correct model control crystals *must* be used, however. The system is compatible with most makes of model control servo.

fig A.1 *underside of printed circuit layout for the transmitter*

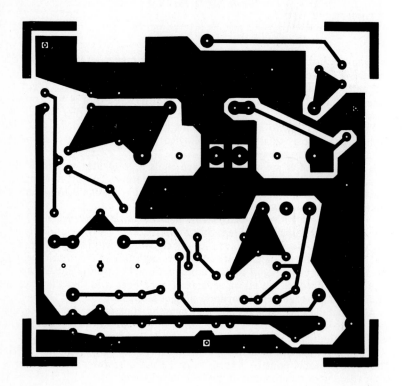

fig A.2   *component positions for the transmitter; TR₁ should be fitted*
with a clip-on heat radiator; the connecting wire to the aerial
should be short; a 1.4 m telescopic aerial is recommended

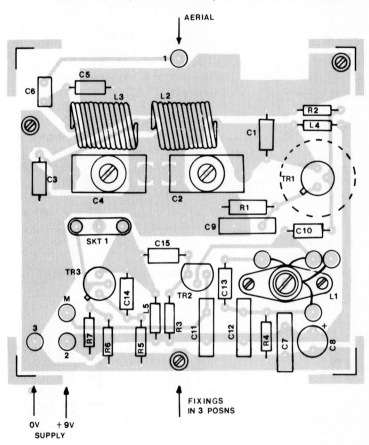

Transmitter component layout

R_1 : 2.7	C_1 : 47p	C_{12} : 0.047μ	TR1 : BFY51
R_2 : 100	C_2 : trimmer, 40p	C_{13} : 27p	TR2 : 2N3702
R_3 : 2.2k	C_3 : 100p	C_{14} : 10p	TR3 : BC109C
R_4 : 100	C_4 : trimmer, 40p	C_{15} : 27p	$L_1 - L_3$: (see text)
R_5 : 150	C_5 : 47p		L_4 : 15μH
R_6 : 10k	C_6 : 0.01μ		L_5 : 15μH
R_7 : 22k	C_7 : 0.1μ		
	C_8 : 10μ		SKT1 : type 25u socket
	C_9 : 0.047μ		
	C_{10} : 100p		
	C_{11} : 0.047μ		

fig A.3 *underside of printed circuit layout for the transmitter encoder*

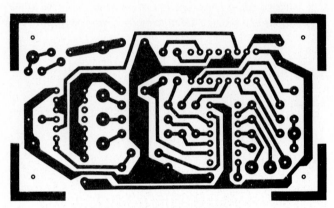

Printed circuit board for encoder.

fig A.4 *component positions for the transmitter encoder; link wires are marked 'LK', and the numbered connecting pads go to the control potentiometers; if commercially made 'joystick' controls are used, the values of the fixed resistors R_1-R_7 may need to be changed.*

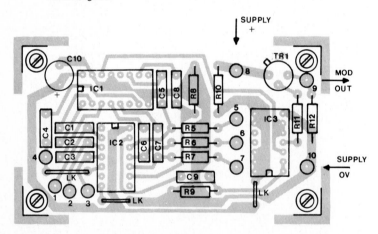

Encoder component layout

R_1–R_4 : (see text)	C_1–C_7 : 0.022 μ	TR1 : BC479 (or BC179)
R_5 : 68k	C_8 : 0.01 μ	IC1 : 4022
R_6 : 68k	C_9 : 0.0047 μ	IC2 : 4078
R_7 : 68k	C_{10} : 10 μ	IC3 : 4001
R_8 : 1.2M		
R_9 : 120k		
R_{10} : 220		
R_{11} : 10k		
R_{12} : 4.7k		

fig A.5 *underside of printed circuit layout for the receiver*

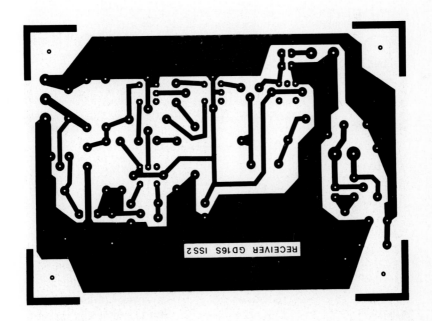

fig A.6 *component positions for the receiver; 'LK' is a wire link*

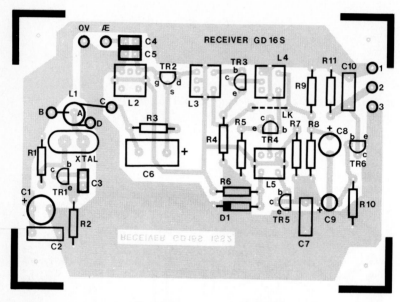

Receiver component layout

R_1 : 100k	C_1 : 47μ	D1 : 1N4148
R_2 : 1.5k	C_2 : 22n	TR1 : 2N4124*
R_3 : 1.5k	C_3 : 15p	TR2 : 2N5457
R_4 : 47	C_4 : 15p	TR3-6 : 2N4124*
R_5 : 47	C_5 : 47p	L_1 : (see text)
R_6 : 15k	C_6 : 470n	L_2 : Toko 113CN 2K159DZ
R_7 : 1k	C_7 : 22n	L_3 : Toko LPC 4200A
R_8 : 150k	C_8 : 10μ	L_4 : Toko LPCS 4201A
R_9 : 3.9M	C_9 : 2.2μ	L_5 : Toko LMC 4202A
R_{10} : 33k	C_{10} : 22n	SKT1 : type 25u socket
R_{11} : 4.7k		

All electrolytic capacitors
should be tantalum 'bead' type,
10V working or more. All resistors
0.25W, 10%.

*BC184L can be used instead
of 2N4124 if more easily
available.

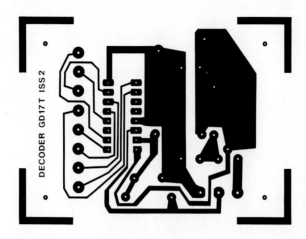

fig A.7 *underside of printed circuit layout for receiver decoder*

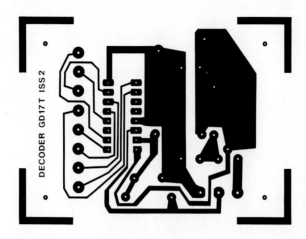

fig A.8 *component positions for the decoder. The numbered connections go to the servos, along with suitable power supply lines taken direct from the battery. It is advisable to use a socket for the CMOS IC.*

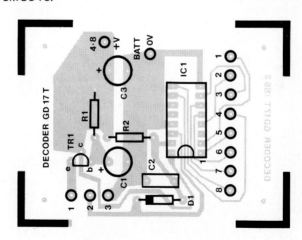

Decoder component layout

R_1 : 1.5k	C_1 : 22μ	ICI: 4017
R_2 : 470k	C_2 : 10n	D1 : 1N4148
	C_3 : 47μ	TR1 : 2N4124 or BC184L
		Battery : 4.8V NiCd

384

fig A.9 *the output waveform of the transmitter encoder; all channels are set about midway; the long gap is the synchronising period*

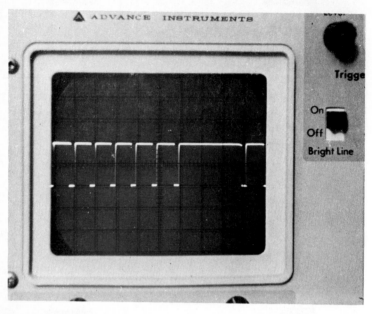

fig A.10 *the waveform at the receiver output*

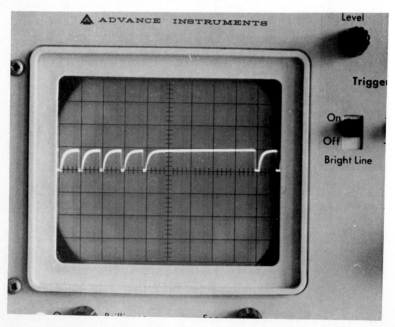

fig A.11   *the receiver decoder output, measured at the RESET input of the 4017, pin 15; the large excursion is the only time that the signal should rise above half the supply voltage, to reset the counter*

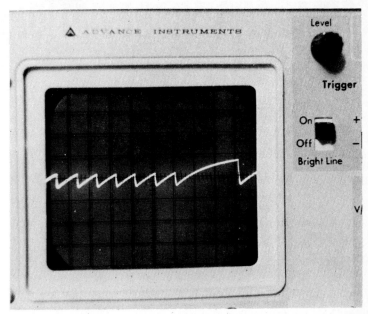

NOTES ON TUNING THE RADIO CONTROL RECEIVER

When setting up the receiver you will find that L2, L3 and L4 have a powerful effect, and need to be tuned very carefully. L5 has less effect, and is best adjusted to give the longest range and freedom from servo 'jitter'.

As an alternative to setting up using a meter, an oscilloscope can be used to view the output at the collector of TR4, setting L2, L3 and L4 for the greatest amplitude. L5 is still best adjusted as above, but it can be interesting to view the waveform at the collector of TR5.

I would like to express my sincere thanks to Terry Tippett of Micron Radio Control for permission to use the receiver circuits.

GLOSSARY OF TECHNICAL
TERMS AND ABBREVIATIONS

A   Ampere (amp); the unit of electric current.

a.c.   Alternating current.

AF   Audio frequency.

alternating current   An electric current that alternates in its direction of flow. The *frequency* of alternation is given in *herz*.

AM   Amplitude modulation.

amplitude modulation   A system of modulating a carrier in which the amplitude of the carrier is changed in sympathy with the modulating signal.

analogue   A system in which changing values are represented by a continuously variable electrical signal.

astable   A circuit which has no stable condition, and oscillates at a frequency determined by circuit values. See also *oscillator*.

audio   Relating to a system concerned with frequencies within the range of human hearing.

bandwidth   The range of frequencies to which a system will respond in the required manner.

base   One terminal of a *bipolar transistor*.

BASIC   The most popular computer language for small computers. The letters of the acronym stand for **B**eginners **A**ll-purpose **S**ymbolic **I**nstruction **C**ode.

binary   A number system to the base 2.

bipolar transistor   A transistor in which current is carried through the semiconductor both by holes and electrons.

bistable   A system which can have two stable states, and which can remain in either state indefinitely.

Boolean algebra   A system of formal logic used for minimising complex digital systems.

breakdown   A sudden loss of insulation properties, resulting in a rapid and large current flow. Typically, breakdown might occur in a semiconductor device operated at too high a voltage.

bus   A group of wires having a related function. Used particularly in computers.

candle   Unit of luminous intensity.

capacitor   A component used in electronic circuits, exhibiting the property of capacitance.

CdS Cadmium sulphide; used in photoresistors.

chrominance In a television system, the part of the television signal concerned with colour.

class A amplifier An amplifier in which the output transistor is operated at approximately half the supply voltage, resulting in a continuous heavy current flow, but low distortion.

class B amplifier An amplifier in which the output is shared between two transistors, resulting in much more efficient operation but potential problems from crossover distortion

CMOS **C**omplementary **M**etal **O**xide **S**emiconductor. A family of digital logic, featuring medium-speed operation, and very low current requirements.

collector One terminal of a bipolar transistor.

conductor A material through which an electric current can flow relatively easily.

conventional current Electric current, regarded as flowing from positive to negative.

CPU **C**entral **P**rocessor **U**nit. The main number-processing and control section of a computer. In a *microcomputer* the CPU will be a *microprocessor*.

CRT Cathode ray tube.

crystal Usually refers to quartz crystal, used as a precision timing element in many circuits. May refer to a piezoelectric crystal pick-up.

Darlington pair Transistors used in a configuration giving high gain and high input impedance.

dB Decibel: one-tenth of a bel, the unit of relative power. See *decibel*.

d.c. Direct current.

decibel One-tenth of a bel. A measure of power, on a logarithmic scale. Symbol dB. The decibel is a convenient unit for representing a very large range of powers.

demodulation The recovery of a modulating signal from a modulated carrier.

denary The 'normal' number system, to the base 10.

diac A bidirectional breakover diode. Often used for triggering a *triac*.

digital electronics The branch of electronics concerned with the processing of digital systems, usually in *binary*.

DIL-pack The standard package used for *digital* integrated circuits, and many *analogue* integrated circuits.

diode A component, either semiconductor or thermionic, that permits current to flow through it in one direction only.

direct current An electric current that flows steadily in one direction (compare *alternating current*).

discrete Used to refer to systems constructed from individual components—e.g. transistors, capacitors, diodes, resistors—as opposed to systems made using *integrated circuits*.

disk A system of magnetic recording used to store large volumes of information. Used with computers, the disk enables relatively rapid recovery of recorded information, as compared with tapes.

doping The addition of tiny amounts of impurities to semiconductor material during the manufacture of semiconductor devices.

dynamic RAM A **R**andom **A**ccess **M**emory circuit, used in computers, in which information is stored in the form of electric charges. Dynamic RAM must be read at intervals of a few milliseconds if the information is not to be lost. However, dynamic RAM is cheaper than *static RAM* and can be packed more densely.

emitter One terminal of a *bipolar transistor*.

EPROM **E**rasable **P**rogrammable **R**ead-**O**nly **M**emory. A *ROM* in which data are fixed, but can be erased if required by the application of suitable electrical signals or ultraviolet light.

extrinsic semiconductor A semiconductor material produced artificially by the addition of impurities.

F Farad: the unit of capacitance. See *farad*.

fan-in The number of standard devices that can be connected to the input of a digital circuit.

fan-out The number of standard devices that can be connected to the output of a digital circuit.

farad Unit of capacitance. The farad is a very large unit, the largest practical unit being the microfarad.

ferrite A finely divided ferrous dust, suspended in a plastic material. Ferrite has useful magnetic properties, but does not conduct electricity.

fibre-optic A glass or plastic fibre used for the transmission of light over long distances.

field timebase In television, the basic *oscillator* used to control the vertical scanning of the picture.

field-effect transistor A type of transistor characterised by a very high input resistance.

flip-flop General term for a bistable, astable or monostable circuit.

flux Various meanings, but usually a resin added to solder in order to prevent the formation of oxides on the material being soldered.

FM Frequency modulation.

frequency The number of waves, vibrations or cycles of any periodic phenomenon, per second. Unit *hertz*.

frequency response Generally the range of frequencies that can be processed by an electronic system.

gain The factor by which the output of a system exceeds the input.

gate (a) A component in digital logic circuits.
　　　(b) One terminal of a field-effect transistor, or other semiconductor device.

Ge Chemical symbol for germanium, a semiconductor.

H Henry: the unit of inductance.

Hall-effect A change in the way that current flows through a conductor or semiconductor when subjected to a magnetic field.

henry Unit of inductance.

hertz The unit of frequency. One hertz equals one cycle per second.

hexadecimal A number system to the base 16—commonly used in computing.

hi-fi High-fidelity—used to apply to audio systems that reproduce the entire audio spectrum, and beyond, with minimal distortion.

Hz Hertz: the unit of frequency.

IC Integrated circuit.

IGFET Insulated Gate Field-Effect Transistor.

impedance The ratio of the voltage applied to a circuit to the current flowing in the circuit. Similar to *resistance*, but applicable to alternating currents and voltages.

inductor A component exhibiting inductance.

insulator A material through which electric current will not easily flow.

integrated circuit An electronic system, or part of a system, produced on a silicon chip using microelectronic techniques.

intermediate frequency In radio and television, the frequency generated as a result of mixing the local oscillator and incoming signal.

JUGFET JUnction Gate Field-Effect Transistor.

Karnaugh mapping A visual technique used in the planning of digital systems for the minimisation of logic circuits.

LC oscillator An oscillator using an inductor and a capacitor in a resonant circuit as a timing element.

LCD Liquid crystal display. See *liquid crystal display*.

LED Light-emitting diode.

light-emitting diode (LED) An electronic component in which electric current is converted directly into visible or infra-red light.

line timebase In a television, the *oscillator* circuits concerned with horizontal scanning of the picture.

linear electronics Electronic systems in which quantities are represented by continuously varying electrical signals. See also *digital electronics, analogue*.

liquid crystal display (LCD) A reflective display, used in digital systems for the presentation of output. The liquid crystal display is characterised by a very low power consumption.

logic Usually used as an abbreviation for 'digital logic', referring to systems involving logic gates.

luminance In television, the part of the signal concerned with the brightness of the image on the tube.

microcomputer A computer in which the *CPU* is a *microprocessor*.

microprocessor A computer *CPU* constructed using large-scale integration in which all the *CPU* circuits are fitted into a single integrated circuit.

modulation Variation of the frequency, phase or magnitude of a high frequency waveform in accordance with a waveform of lower frequency.

monostable A system with a single stable state.

MOS Metal Oxide Semiconductor.

MOSFET Metal Oxide Semiconductor Field-Effect Transistor.

multimeter A general-purpose measuring instrument, usually able to measure resistance, current and voltage.

negative feedback Feedback applied to a system in such a way that it tends to reduce the input signal that results in the feedback.

NiCd Chemical symbols for nickel and cadmium; used to refer to nickel–cadmium accumulators.

NMOS *n*-channel MOS.

npn Negative–positive–negative (although always pronounced 'en–pea–en'); refers to one of the two alternative types of bipolar transistor.

NTSC National Television Standards Committee. The American body that defined the American television standard. 'NTSC' is used to refer to the type of TV system used in the USA.

operational amplifier A highly stable, high gain, d.c. amplifier, usually produced as a single integrated circuit.

optoelectronics Electronic systems or devices that involve the use of light.

opto-isolator An *optoelectronic* component used to couple signals from one system to another, while retaining a very large degree of electrical isolation between the two systems.

oscillator An electronic system that produces a regular periodic output.

oscilloscope An instrument for displaying electrical waveforms on a cathode ray tube.

Ω Ohm: the unit of resistance.

PAL Phase Alternation by Line. The colour television system used in the UK and elsewhere. It has advantages over the *NTSC* system that preceded it.

passive component A component that does not involve the control of electrons in a thermionic or semiconductor device.

PCB Printed circuit board.

photoresistor Also known as an LDR (light-dependent resistor). A resistor whose value depends upon the amount of light falling on it.

piezoelectric effect The direct conversion of electrical to mechanical energy, or vice versa, in some crystalline materials.

PMOS *p*-channel MOS.

pnp positive–negative–positive (although always pronounced 'pea–en–pea'); refers to one of the two alternative types of bipolar transistor.

positive feedback Feedback applied to a system in such a way that the feedback tends to increase the input signal causing the feedback.

potentiometer A variable resistor having connections to each end of the track and also to the brush.

PPM Pulse-position modulation.

program A set of instructions used by a computer.

PWM Pulse-width modulation.

quartz crystal oscillator A very stable oscillator, depending for its stability on the electromechanical properties of a quartz crystal.

RAM Random Access Memory. The main storage area in a computer system.

raster The pattern of horizontal lines produced on a television screen.

relay An electromechanical device in which an electric current closes a switch.

resistance The property of a material that resists the flow of electrical current.

resistor A component exhibiting a known amount of resistance.

RF Radio frequency.

ROM Read Only Memory. Used in computer systems to store programs and information that is not lost when the system is switched off.

Rx Abbreviation for 'receiver'.

semiconductor A material with properties that lie between those of *insulators* and *conductors*. Extensively used in modern electronics.

Si Chemical symbol for silicon, a semiconductor.

speaker (loudspeaker) An electromechanical device for converting electrical energy into sound.

static RAM Random Access Memory in which information is stored in bistable devices. See *dynamic RAM*.

superheterodyne A radio receiver system in which the radio frequency input is mixed with a frequency generated within the receiver to produce an *intermediate frequency*.

teletext Any system that involves production of digitally generated text and pictures using standard television systems.

thermionic Electronic devices involving electrons generated by heat, usually in a vacuum.

thyristor A component similar to a semiconductor diode but having in addition a *gate* connection by which the component, normally non-conducting, can be triggered into conduction.

tolerance Generally the amount by which a specified component value can vary from the marked value.

triac A semiconductor component similar to the *thyristor* but which will conduct in either direction.

TTL Transistor–transistor logic: one of the two major 'families' of digital logic circuits. Uses more power than CMOS, but is capable of higher operating speeds.

Tx Abbreviation for 'transmitter'.

ultrasonic A frequency above the range of human hearing. Note that 'supersonic' is now generally used to mean 'faster than the speed of sound'.

unijunction transistor A semiconductor device used in some *oscillators*.

V Volt: the unit of electrical potential.

varicap diode A semiconductor diode in which the junction capacitance varies according to the applied voltage. This effect is inherent in all semiconductor diodes, but in the varicap diode the property is deliberately enhanced. Used in tuning circuits in radio and television.

VDU Visual display unit.

video (a) In television, the demodulated vision signal.

(b) More generally, anything relating to the recording, replaying, transmission or reception of pictures.

wavelength The physical distance between two similar and successive points on an alternating wave.

Zener diode A semiconductor diode, used for voltage regulation. When The Zener diode is reverse-biased, it exhibits a sudden increase in conductivity at a certain specific voltage.

RECOMMENDED FURTHER READING

The following books may all be of interest; books marked with an asterisk*
are suitable for more advanced studies:

L. Basford, *Electricity Made Simple* (W. H. Allen, 1968).
Graham Bishop, *Linear Electronic Circuit and Systems*, 2nd edn (Macmillan, 1983).
J. Cluley, *Electrical Drawing I* (Macmillan, 1979).
P. E. Gosling, *Beginning BASIC* (Macmillan, 1977).
*P. E. Gosling, *Continuing BASIC* (Macmillan, 1980).
P. E. Gosling, *Mastering Computer Programming* (Macmillan, 1982).
G. King, *Beginner's Guide to Radio* (Newnes–Butterworth, 1973).
R. Lewis, *Electronics Servicing: Theory* (Macmillan, 1981).
R. Lewis, *Electronics Servicing: Core Studies* (Macmillan, 1983).
*J. Millman, *Microelectronics—Digital and Analog Circuits and Systems* (McGraw-Hill, 1979).
N. M. Morris, *Digital Electronic Circuits and Systems* (Macmillan, 1974).
*N. M. Morris, *Semiconductor Devices* (Macmillan, 1976).
G. Wright, *Mastering Computers* (Macmillan, 1982).
*J. Zarach and N. M. Morris, *Television—Principles and Practice* (Macmillan, 1979).

INDEX